Oil,
The 4th Renewable Resource

Shawn Alli

Background cover art by Tuomas Pekkarinen

Printed in the USA. 2nd edition, November 2013.
Distributed by Ingram Content Group.

ISBN 978-0-9917182-0-7

Shawn Alli
Suite # 418
2 Toronto St.
Toronto, Ontario, M5C 2B5
Canada

www.shawnalli.com

Library and Archives Canada Cataloguing in Publication

Alli, Shawn, 1982-

Oil, the 4th renewable resource / Shawn Alli.

Includes bibliographical references and index.
ISBN 978-0-9917182-0-7

1. Deep-earth gas theory. 2. Hydrocarbons. 3. Petroleum industry and trade. 4. Petroleum--Prices. 5. Energy policy. I. Title.

TN870.5.A45 2012 553.2'801 C2012-906931-0

Dedicated to the Aboriginal youth

of Pehdzeh Ki and Lutsel K'e

that continue to help me

on my path.

Table of Contents

Introduction

Oil is the second most valuable and abundant commodity after water. Gold, diamonds, silver and copper don't compare to the prestigious title oil enjoys. Empires fall because of it while others become wealthy beyond any scope of measurement. People rejoice in finding it and will kill to keep it flowing. But the true origin of oil is one of suppression and censorship, one that's too incredible to believe—that oil is a renewable resource along with water, solar and wind. It's more accurate to say "Hydrocarbons, the 4th Renewable Resource." But not everyone knows what hydrocarbons are so I'll stick with oil. These are words no environmental scientist will speak of. Why not? Because this will undermine the entire green climate change movement and cause it to lose credibility in the eyes of the unforgiving public, who grudgingly go along with recycling and composting so they can feel they're helping the Earth. Oil corporations can't openly admit oil into the renewable resources category because of the profit factor. It's just not good business to purposely lower the value of your commodity for the benefit of all.

This book will allow any reader to gain a solid understanding of the deception and manipulation of oil as a fossil fuel. Including this book, there are now five whole books circulating in North America that advocate the abiotic/abiogenic theory of hydrocarbons such as oil, natural gas and coal. The abiotic theory advocates that hydrocarbons are the result of non-biological primordial elements that are naturally in the earth. Abiotic theorists don't dispute the fact that biological molecules are in oil, but rather the explanation of how it gets there. It's not that biological material magically creates oil, but rather living microbial life contaminates oil as it rises upward to the surface of the earth.[1]

This book departs from its predecessors by taking a more practical approach with humorous graphics. The average individual, high school, and university student will be able to understand all of the information with clarity. This book is also strong enough to cause university professors of geology, earth science and environmental studies to question their beliefs in the formation of oil.

What qualifications enable me to go where others fear not? I don't have a PhD nor am I a scientist or geologist, so what right do I have to say otherwise and why should you believe me? A valid question. The simple answer is my social status. I'm a philosopher, which means I have no significant social status in the Western world. Women don't look at us with eyes that hunger for more depth to understand our perceptual framework. If Socrates, Plato and Aristotle can see what philosophy has become in this era they'll gladly choose death by poison than to exist in this time.

But the monumental drop in social status for my intellectual skills is exactly what allows me to write this book. It can't be done by any individual that has a title, be it a doctor, a scientist, a lawyer, a celebrity or a politician. Such individuals have too much to lose when the complaint letters, angry citizens and mainstream media ridicule begins to swallow another victim. Being a philosophical researcher I have nothing to lose, no title, no tenure, no significant status in this world or the next. This is the type of character that's necessary to question and uncover truth. If more philosophers leave the university faculty positions and start taking up roles in active citizenship, emotional and intellectual growth will become the driving force of the world. Until that time arrives you have me. And I'll ask you to hold your judgments until the end of the book. After that, let it all out and hopefully you'll see the world through a new filter of perception.

One last thing to keep in mind is the writing style of this book. All text is in the present tense with all past events in the historical present as well. Newspapers and some fictional novels use the historical present to a small degree. Utilizing this literary technique for an entire non-fictional text is rare; but going against convention is also the purpose of this book.

Chapter 1:

The Myth of Oil as a Fossil Fuel

The idea that hydrocarbons on earth are the chemical remains of surface life that has long been buried and pressure cooked into petroleum and natural gas has been exceedingly difficult to unseat. I have been trying to do so since 1977.

Thomas Gold, The Deep Hot Biosphere: The Myth of Fossil Fuels.

Most individuals think of oil as a source of fuel, yet in the beginning it's much more disturbing. I'm not talking about oil as a lubricant or a source of light, that happens as well, but the larger shock is finding out its use as a cure-all medicinal medicine. In the 1850's individuals regularly use crude petroleum oil as a medicinal cure for liver problems and bronchitis—ingesting it!!![1] Personally, as a philosopher I may like to think that my philosophical constructs are "way out there," but ingesting crude oil definitely beats it out by far. Luckily, time enables scientists and engineers to find better uses for it, such as fuel. Despite all the negativity and stigma in using oil, the industrial revolution is one of the greatest accomplishments in creating modern society, yet none of it can happen without oil and coal. But the social costs in using oil and coal as a catalyst for modern society, is the fear of scarcity and peak oil. The myth of oil scarcity may have its modern roots in corporate monopolies, but that alone isn't a strong enough starting point to enable the deception of the global general public. A foundation of scarcity is necessary, and that's where the term "fossil fuel" comes into play. Take away this foundation and everything crumbles.

Turning to my old friend collecting dust on the bookshelf I tear open my Merriam-Webster's dictionary and find the term fossil fuel as "a fuel formed in the earth from plant or animal remains."[2] There's no surprise in this definition, but the 1835 origin is of interest. I decide to contact Merriam-Webster's for clarification, and amazingly they're only too happy to give me the references. One day of waiting for the email is all that's necessary. Sadly, I doubt that dictionary publishers get many requests like this and are waiting for the general public to finally make use of them. Their response points me to an 1835 text by J. Holland, *The History and Description of Fossil Fuel, the Colliers, and the Coal Trade of Great Britain.* Clearly, the term is in full usage in this early part of the 19th century; but is this the real beginning?

Internet searches for the origin of fossil fuels push the date back to 1757 by the Russian scientist Mikhail Vasilyevich Lomonosov. The scientific community recognizes him for his astronomical works in theorizing that Venus has an atmosphere. Lomonosov is a respectable pre-industrial scientist in his own time, and scientists today still see him as a scientific genius. But is it correct to attribute the definition of fossil fuels to him? This is where it gets tricky. If anyone does an internet search on the origins of oil they'll come up with a quote by Lomonosov stating that it originates from tiny bodies of animals in sediment. Alright, where's the source? What text? Or is it a speech? And that's the problem; there is no text or reference. The only information available on the internet says these words are from the Russian Academy of Science in 1757. Contacting the Saint Petersburg Scientific Center, no such record exists. It's quite possible the text or document no longer exists through numerous political and economic transitions; but if there are no records there's no proof, only speculation. The Center also doesn't have his 1763 publication *First Principles of Metallurgy or Mining* (English translation) that supposedly has a chapter, *On the Earth Layers,* dealing with the origin of oil. On the internet Lomonosov is the father of the definition of fossil fuels. Trade books, media articles and university lectures also reference him in regards to the origin of fossil fuels.[3-4] But the fact is...there's no proof. Perhaps he does make a speech in 1757 about oil. And his 1763 publication is not in doubt for its existence, but it's not available for the general public to scrutinize. This means that Lomonosov is not credible enough to be the originator of the term fossil fuels. If an individual does have the 1757 Russian text lying around and the Saint Petersburg Scientific Center verifies it, I'll gladly concede. Until then, the distinction goes to the German/Polish chemist Casper Neumann.

In his 1759 publication, *The Chemical Works of Casper Neumann*, Casper describes the use of coal rather than giving an actual definition:

> Note: the (s) substitutes for (f) for readability.
>
> *Pit coal, Carbo fossilis, Anthrax, Lithanbrax, is a bituminous fossil found in several parts of the world, more plentifully in England and Scotland, from whence it is exported, in large quantities, into other countries, for the common purpose of fuel.*[5]

The global general public believes that oil is the lifeblood of the industrial revolution, but that's not exactly true. With large quantities of exports common in the 1750's, it's coal that gets that distinction. Although there are no reliable statistics, Western society mines coal for over 250 years and still going strong. To think that we're going to run out is merely due to a narrow understanding of the natural cycles of the Earth. The key point to remember about the origin of fossil fuels is that there's no deliberate deception or malicious intent to fool the public. Natural science in the 18th century is very different from the modern scientific method. It's an era that advocates intelligent explanations from observing the natural world:

> *One reason was the belief in earlier times that hydrocarbons were specifically organic substances: hence the name 'organic carbon' for all forms of unoxidized carbon. The knowledge that hydrocarbon are abundant in the universe, and on many of the other planetary bodies of our solar system, was not available at the time.*[6]

Aside from the lack of information about nature through laboratory tests and verifiable experiments, natural science in the 18th century is a naive but sincere attempt at understanding the planet. Underneath all of the methods, processes and actions, it's ideology that shapes all discoveries:

> *In earlier times there was the belief that the earth had formed as a hot, molten body. In that case no hydrocarbons or hydrogen would have survived against oxidation, nor would any of these substances have been maintained in the interior after solidification. With that belief, there seemed no other possibility of*

accounting for the hydrocarbons embedded in the crust than by the outgassing of carbon in the form of CO_2, produced by material that could have survived in a hot earth, and subsequent photosynthesis by plants that converted the CO_2 into unoxidized carbon compounds.[7]

Even though you're now aware of the origins of fossil fuels, if you ask the general public, teachers, scientists, and university environmental professors where the term originates from, you'll get a pondering facial expression. But this lack of awareness of a front runner championing the term is not due to coincidence but rather an intentional silence in the modern era; and this is an important distinction. Intentional deception isn't the purpose for the original definition and description of fossil fuels, but rather, it seeks to study and understand the natural laws of the Earth. This is in sharp contrast to the modern definition where the myth of scarcity propagates intentionally. However, thousands of modern researchers and authors who advocate oil as the past remains of vegetation and animals don't want to include the "D" word, and that of course is "Dinosaurs." Claims in print books and publicly on the record, that oil comes from dinosaur remains, invites a chuckling ridicule from the abiotic theorists *(the term abiotic will represent both abiotic and abiogenic terms).* Not wanting their conversations and claims to sound unintelligent, peak oil theorists substitute the word dinosaurs for more acceptable terms such as *vegetation, algae, plankton and ancient forestry.* While all dictionaries and encyclopedias define fossil fuels as the same, the term "animal remains" is no longer welcome in its spot beside vegetation.

PEAK OIL CONVENTION
Vegatation & Animal Remains Booth
Vegatation only booth
It says animal remains, not dinosaurs!!!
I never feel welcome in these conventions.

It's not that "animal remains," is no longer part of the definition. Ask any mainstream, corporate or research scientist what a fossil fuel is and they'll use the term without hesitation. But ask any peak oil theorist or author against the abiotic theory, and the term animal remains usually falls into a dark silence. The peak oil theory states that the supply of oil has a peak and when that occurs the supplies have to drop (more in Chapter 2). In January 2006 a radio debate takes place between Dr. Jerome Corsi and Michael Ruppert on Coast to Coast AM (a late night radio talk show for alternative media). Mr. Ruppert argues for the peak oil theory while Dr. Corsi argues for the abiotic oil theory. The mediator (George Noorey) inquiries into where oil comes from if not from dinosaurs. Mr. Ruppert's response is:

> *The oil and natural gas come from organic matter, carbon based life forms...*
>
> *nobody says that oil came from dinosaurs, that is a common myth, nobody has argued and I'm not arguing that oil came from dinosaurs, we gotta stop that...what it is, is an accumulation of algae and plankton which the oceans were filled with.*[8]

I may not have a PhD but what constitutes organic matter is all life: plants, bacteria, microbes, dinosaurs, animals and humans. To say that one is shut out because it sounds "unintelligent" isn't fair to rational scientific inquiry. But even if you're in a gentle mood and let them exclude those big ol' misunderstood dinosaurs, and claim oil comes from algae and plankton, the question remains, how much algae and plankton are necessary to produce trillions of barrels of oil? Confusing looks and silent shoulder shrugs will be the usual reaction.

The disassociation of dinosaurs from the peak oil theory is one of the most blatant forms of logical fallacy. To speak of oil as decaying vegetation and not decaying dinosaur flesh is a crime that all peak oil theorists should be accountable for. They should publicly admit their claims represent both so they can feel the ridicule as the general public begins to chuckle. Such an event only occurs in my visualizations. But we're all free to imagine anything, so there's no harm adding a few whippings and a dunce cap in the scenario.

Oily Coffins?

If all organic matter is up for grabs as it transforms into oil, natural gas and coal then what happens to our loved ones when they pass away and are six feet under in the cemetery? This is another issue that biogenic and peak oil theorists don't like to bring up. When we pass away do we turn into oil? Abiotic theorists will reply with a chuckling no. Biogenic theorists (advocating the creation of hydrocarbons from biological material), will have the same reply with an irritable look on their face. And technically speaking, they're both correct. But let's indulge our imagination with a hypothetical scenario:

100 million people fall victim to a horrible disease and die. The surviving humans pile the bodies in a sedimentary grave with thousands of small bacterial life flourishing below and above the surface. They cover the bodies with soft muddy microbial earth. 100 million years pass and the bodies sink one kilometer deep. The future humans find out what happens and goes to the muddy area and dig one kilometer deep. What will they find?

The abiotic theorists will say only skeletal remains will greet our future humans. The biogenic and peak oil theorists will say the bodies are now hydrocarbons such as oil, natural gas or coal and will be ready for oil refineries. The reason why they have to admit this is because the biogenic process forces them to. Sufficient temperature, passage of time, pressure, and organic life are all that's necessary to produce hydrocarbons according to the theory.

Looking at the biogenic theory from this point of view causes the average individual to have a quizzical expression on their forehead, *okay...that's ah...interesting*. If you ask any individual on the street whether they believe their loved ones will turn into oil with sufficient time, pressure and temperature the results will be a dismal no.

Professor Thomas Gold, an American astrophysicist who advocates the abiotic theory, explains why living creatures that pass away don't gain the atoms necessary to formulate hydrocarbons:

> *Biological debris buried in the earth would be quite unlikely to lose oxygen atoms and to acquire hydrogen atoms in their stead. If anything, slow chemical processing in geological settings should lead to further oxygen gain and thus hydrogen loss....How then, could biological molecules somehow acquire hydrogen atoms while, presumably, degrading into petroleum?* [9]

This is what bothers me about science class in elementary schools. On the one hand teachers are saying dinosaurs decompose over time and their bones are left behind. On the other hand oil comes from the rotting flesh of dinosaurs as it transforms from the heat and pressure. The deepest dinosaur is about two kilometers deep under the seabed with a date of 200 million years,[10] yet oil corporations find oil at depths of eight kilometers or more. It doesn't make sense. This is one reason why I can never go into teaching. What a student learns in the classroom and what it's really like out of the classroom is far too wide a gap. Being a philosopher I prefer not to instill young minds with a false reality—telling them delusional stories from curriculum textbooks with contradictions and propaganda. I'm not that devious.

But scientific validation isn't necessary to prove that people or any living creature turns into oil. The empirical evidence is all around us. Any corpse that exists for thousands of years has to show some indication it's in the process of decaying into oil, right? The best place to look is at the local museum, in the ancient Egypt exhibit. Mummies that are 5000 years old still have soft tissue on them due to excellent preservation at the time. When does the oil start gushing out? Even placing these mummies into a deep hole and leaving it for millions of years won't result in any oil coming up.

Perhaps mummies are too complex; a simple example is to observe a plant. Just look in your backyard, find a weed, a flower or grass and pull it out. Over the next few days the plant will start to rot. Why? Because it's not able to absorb carbon dioxide and nitrogen from the atmosphere. It doesn't miraculously gain new hydrogen atoms to become a hydrocarbon. Hydrocarbons can be any compound that has short or long chains of hydrogen and carbon molecules. Oil, natural gas and coal are examples of hydrocarbons. What about water? Water has one oxygen and two hydrogen atoms. Can a decaying plant retrieve these hydrogen atoms and begin the oily process? Nope, it'll just sit there looking at you wanting to say, *it's not going to happen.*

So why do students, the general public, and scientists believe this if it doesn't make sense? The simple answer is repetition. If someone hears a claim about the creation of oil over and over again, the end result is a firm belief in the mind. But how is this happening in the present? What are the exact actions allowing the myth of fossil fuels and scarcity to continue? Tragically, the mechanisms are part of every aspect of our lives, and no one is exempt from it.

Chapter 2:

Propagating the Peak Oil Theory Today

World oil production can probably keep going up for another six or eight years. But some time in the 1980s it can't go up much more. Demand will overtake production. We have no choice about that.

US President Jimmy Carter's Energy Policy Speech, April 18, 1977.

With only 2% of the world's oil reserves, we can't just drill our way to lower gas prices...Not when we consume 20% of the world's oil.

US President Barack Obama's Weekly Address, March 3, 2012.

The myth of fossil fuels and scarcity go hand in hand. Even though the theory of fossil fuels begins in 1759, the theory of scarcity is a modern artificial creation. The myth of peak oil and scarcity can only propagate with a finite supply of fossil fuels. Even with that, a strong speaker is still necessary to entice the public. And the office of the US presidency is quite capable of fulfilling this role. In President Jimmy Carter's energy speech he emphasizes the threat of scarcity, suggesting the end of the current western lifestyle.[1] He even promotes ten economic energy principles with good intentions, yet none are put into practice in his time or 30 years later. Under President George W. Bush the price of oil reaches $150.00 a barrel with massive failures in his energy policy.[2]

And this era is no different as Barack Obama claims the US holds only 2% of proven reserves and consumes 20%.[3] According to the US Energy Information Administration (EIA) the 20% consumption is correct, yet the 2% of proven reserves isn't. Doing the calculations of US oil production in comparison to the total world oil supply, it equals 11.6%. While it's a significant difference it's actually not important. The reason is because the message is the potent force. By claiming the 2% reserves and 20% consumption, it's similar to Carter's message—that we can't sustain our current lifestyle—that there won't be enough for everyone. This is the same Malthusian concept of scarcity the general public hears on a daily basis. So what's the government's alternative? Nothing really. While there are commitments to invest in more renewable resources, the results are nowhere to be seen. Are there automobile manufacturers running their operations entirely on solar power? If there are, they operate in the shadows. Is the airline or weapons industry running on solar or wind energy? Are cars running on a fuel different than hydrocarbons? Are contractors building houses on a large scale that are self-sustainable and require no electricity from the grid? The simple answer is no. As great as these goals are they're pipe dreams that Western governments are selling to the general public and to school kids. And this is the heart of the matter. The most effective means to convince a population of the scarcity myth is to start when they're young, the education system.

Schools

The school system is the best mechanism to instill a concept into a child's mind; after all, school is a part of everyone's lives for 12-16 years. And who controls the core of the education system? The government. Sure, there are private schools that have their own curriculum, but the public as a whole can't afford that. And any school that gets funding from the government has a list of textbooks that can be in the classroom. If a textbook is not on the list and someone reports it, either government funding ends or in most cases, the administrator throws the book out. Advocating the myth of scarcity through the fossil fuel theory is the best way to condition students to accept the myth without question.

Being a Canadian citizen I'm not able to locate all science textbooks in the US and European school system. In the US local jurisdictions have control over textbooks in the curriculum; while in Canada, it's a provincial matter and much easier to find the actual texts. I'll list the textbooks that advocate the fossil fuel and scarcity myth in Canada. I'll be thankful if my American and European counterparts can take up the challenge of researching it in their countries as well.

Canadian science textbooks in government schools

1. *Conservation of Energy*. Grade 5:

> *Most cars, trucks, boats, and airplanes are fuelled by gas and oil, both of which are fossil fuels.*[4]

2. *Physical Geography 7*. Grade 7:

> *Formed from the remains of plants and animals buried underground.*[5]

3. *Experience Canada: A Geography*. Grade 9:

> *About 25 per cent of Canada's electricity comes from burning non-renewable fossil fuels—coal, oil, or natural gas, which contain the stored carbon of once-living things.*[6]
>
> *Since the 1970's conventional drilled wells have been unable to keep pace with our energy consumption. Canada's petroleum industry has scrambled to find new oil sources.*[7]

As a Canadian, I don't recall any energy crisis where there's no energy to generate electricity, drive vehicles, or scrambling to find oil. The 1973 oil embargo increases the price of oil, it's doesn't stop the supply. This quotation is clearly advocating the myth of scarcity—that there won't be enough for everyone.

4. *Earth Matters: Studies in Physical Geography*. Grade 11:

> *Natural gas, for example, was created millions of years ago from decomposing swamp vegetation. While these plants were alive, they stored solar energy in the form of carbohydrates, a substance made through the chemical process of photosynthesis. After the plants broke it down, the energy was stored until we burn it as fuel. Oil and coal are other fossil fuels that were created in a similar way and which also release greenhouse gases when they are burned.*[8]

The massive error in this quotation, which most authors make in regards to fossil fuel creation, is the photosynthetic process playing any part in the creation of hydrocarbons.[9]

5. *Earth Science 9th edition.* Grade 11:

> *By contrast, many other basic resources are classified as nonrenewable. Important metals such as iron, aluminum, and copper fall into this category, as do our most important fuels: oil, natural gas, and coal. Although these and other resources continue to form, the processes that create them are so slow that significant deposits take millions of years to accumulate.*[10]

It's not that the process takes a large amount of time, but rather it starts out billions of years ago and continues every day from that beginning. The Earth is always producing hydrocarbons and will keep doing so until the end (more on this in Chapter 3).

6. *Techno-fix: Why Technology Won't Save Us or the Environment.* University level:

> *For more than a century, human have been burning fossil fuels from the deposits that required millions of year to accumulate, creating a sever imbalance in the global carbon cycle.*[11]

I can only prove without a doubt that the myth of fossil fuels is in full swing in government schools in Canada between grades 5 to university; but I speculate it's happening much earlier for all North American and European schools.

But here's a quirky task you can give to your teacher. The next time your science teacher or professor is talking about hydrocarbons, oil, or fossil fuels, ask them the infamous question: how many dinosaurs does it take to make a barrel of oil? Apart from being a sarcastic quip, this is a serious question that no one has an accurate answer for. The answer eludes me as well. How many dinosaurs and vegetation have to exist millions of years ago in order to equal the amount of oil in production (supply) since the 1860's to the present? Unfortunately I don't have accurate statistics for oil production that goes that far back in history. But I do have reliable statistics from the EIA from 1980 to 2011.

Simple math will give you the numbers to ask your science teacher or professor:

World Oil Supply (in barrels per day):[12]

Year	Quantity
1980	63,987,000
1981	60,602,000
1982	58,097,000
1983	57,927,000
1984	59,563,000
1985	59,156,000
1986	61,533,000
1987	62,099,000
1988	64,394,000
1989	65,518,000
1990	66,435,000
1991	66,339,000
1992	66,552,000
1993	67,101,000

1994	68,636,000
1995	70,304,000
1996	71,986,000
1997	74,219,000
1998	75,680,000
1999	74,838,000
2000	77,708,000
2001	77,640,000
2002	77,041,000
2003	79,496,000
2004	82,981,000
2005	84,417,000
2006	84,403,000
2007	84,249,000
2008	85,306,000
2009	84,328,000
2010	86,788,000
2011	87,109,000

Total 2,306,432,000 barrels of oil per day x 365 days = 841,847,680,000

From 1980 to 2011 the world produces 2.3 billion barrels of oil a day, and multiplying it by 365 days in a year equals 841 billion barrels.

1. How many dinosaurs and vegetation are necessary to produce 841 billion barrels of oil?

2. Taking into account the size of land and sea dinosaurs, how much land mass is necessary for those dinosaurs to exist?

Media

In regards to brainwashing, the mainstream media is second only to school indoctrination about the myth of fossil fuels and the scarcity of oil. Newspaper articles in the past century have numerous headlines such as *Emergency Oil Supply Exhausted,*[13] and *Ten More Years of Oil—What Then.*[14] Oil scarcity articles in the mainstream media enable confusion and hysteria for the general public—manipulating them to drastically alter their lifestyle. I have no problem with an individual choosing a more natural lifestyle (even I aim for that), but to do so out of manipulation and fear is unethical. But when do ethics ever come into the picture for mainstream media? I'm being too harsh of course, but not completely out of line. 1909 is the beginning of the scarcity myth in mainstream media with articles by George Otis Smith, the director of the US Geological Survey (USGS) from 1907-1922 and 1923-1930. In the magazine, *American Review of Reviews*, Smith's subordinate David T. Day writes extensively about the shortage of oil:

> *If such a large average is assumed for the known fields of the United States, a total product of twenty-five billion (25,000,cxx), cxx)) barrels would be expected before the present fields are exhausted.*
>
> *These considerations show that at the present rate of increase in production, supplies of petroleum in the known deposits would be less than the requirements of the trade in the next decade, except in California.*[15]

Smith laments the supply of oil in the 1920 *Munsey's Magazine*:

> *A few months ago I tried to visualize the torrent of oil consumed by the country in 1918 by imagining that it was poured over Niagara Falls. I found that the oil supply of that year equaled the flow of waters from the Great Lakes with their vast*

drainage basin, for three hours and four minutes; but already there is promise that the consumption of oil in the United States in the present year must be likened to flow of Niagara for three hours and four minutes...An estimate of the petroleum resources of the world has just been published by the chief geologist of the United States Geological Survey, Mr. David White, and his figure of sixty billion barrels for the whole world is no doubt exact enough to enable us to see the oil situation in the United States in fairly true perspective.[16]

The 1920's are the worst period for scarcity myths with quick news bites in daily publications:

May 17, 1920 *The New York Times*: "Exhaustion of Our Oil Supply in Twenty Years Predicted."

May 19, 1920 *The New York Times*: "Oil Depletion Here Alarms Officials."

June 20, 1920 *The New York Times*: "Warns of Future Shortage of Oil: Director of Geological Survey Declares That Consumption Has Overtaken Output."

These scares are reflective of the past but also repeat in the late 80's and early 90's. In 1989 *The Economist* magazine article starts off as, "THE United States running out of oil? Apparently: the 35 billion barrels of proven reserves still under its soil will, at the rate oil companies are currently pumping the stuff, last less than ten years."[17] The *Economist* doesn't change its claim the next year either:

THE world is not running out of oil but America is. Consumption and dependence

> *on imports will both soon pass the peaks they reached in 1976...Lobbyists representing everything from conservation to nuclear power are happily predicting a new energy crisis in the 1990s. On current trends, America will have to buy nearly two-thirds of its oil from abroad by the end of the century.*[18]

Working out the calculations according to the EIA, as of 2011 the US government buys about 45% of its oil overseas.[19] A little less than the 66% the article predicts. But why does the US government buy its oil overseas? Well, if you believe oil is a scarce commodity then why use up your own when you can use up someone else's?

In the *2000 Oil and Gas Investor* magazine, the author claims that oil will have to run out eventually but no one really knows when because technology is getting better. Yet oil is definitely going to hit its peak in 2005:

> *Jack Zagar of Malkewicz-Hueni Associates said the world contains an ultimate resource base of 1.8 trillion barrels of conventional oil, 46% of which has already been produced and 91% of which already has been discovered. World production will peak in 2005; its peak discovery rate was in 1965, he said.*[20]

Looking at the EIA statistics, the claims are false. In 2005 the world's oil production is 84 million barrels a day.[21] In 2011 it's 87 million barrels a day.[22] Where's the drop? Did you see it? I do miss some things when I blink my eyelids...an unfortunate annoyance.

The 2002 *EconSouth* article says the real problem isn't supply but refineries to transform the oil into gas, "Therefore, because of capacity constraints, last summer's high oil prices were probably unavoidable."[23] I'm the reaching the boiling point of how many excuses I can tolerate. The author, Robert Eisenbeis, is talking about the 2001 summer gasoline price increases spreading across North America. His justification is that environmental laws and low amounts of refineries push up the prices. Taking into account how oil corporations make a profit, this is bad...why? It's not bad, it's really good. Gouging the general public under the guise of environmental laws is a slap in the face to the environmental green movement. In Chapter 6 you'll learn how the environmental green movement and oil corporations go hand in hand. As for the low amount of refineries in North America, if you have a massive supply of oil and want to increase the price, one solution is to stop building new refineries and run the current ones at capacity. This concept is nothing new and heavily in use by J.D. Rockefeller's Standard Oil Corporation in the 1870's:

> *It is also asserted that a prominent feature of the combination [a monopoly] will be to limit the production of refined petroleum to such amounts as may serve, in a great measure, to do away with the serious periodical depressions in the article. Is it also to be expected that, desiring to curtail the production of refined petroleum in this country, the railroads will not offer any additional facilities for exportation of the crude article.*[24]

Using the environmental movement to ensure federal laws allow only a small amount of refineries to operate is both deviously brilliant and tragically cruel to environmentalists.

One more and I'll stop. In 2004 *The Middle East* magazine does a current affairs article and claims the era of cheap oil is over:

> *For instance, for new sources of oil to be identified and extracted—the price of this important commodity will have to rise—as most of the 'cheap oil' that could be taken out of the ground and sold for $20-30 a barrel is gone. What comes next is a steady increase in the price of oil until the new hikes make it profitable for the producers to extract it from less accessible sources.*[25]

$20-30 a barrel? Unthinkable now, and that's only 8 years ago. Let's do some math and statistics. The average nominal retail price of gasoline at the pumps in 2004 in the US is $1.85/gallon.[26] A barrel of oil costs $36.75 in April 2004[27] and costs a maximum of $109.53 in 2011,[28] that's an increase of three times. Applying this to the price of gas, Americans should be paying $5.55/gallon. As of 2011 the average gasoline price is $3.53/gallon.[29] Something's not right here. If the price of a barrel of oil increases three times, logically, the price has to increase that much to make a profit...right? That's only the case if oil is a commodity on a democratic capitalistic market. The truth is it's not. The price is completely set by oil corporations who spike it with government taxes. In a real market democracy the price at the gas pumps is interdependent on the price of a barrel of oil. The sad truth is that there is no correlation. Oil corporations charge what they want and set prices at their whim while the mainstream media provides the lubricant to justify the prices to the public.

One thing for certain is that mainstream media is overly passionate about having an energy crisis. Almost every article has the implications of an energy crisis where the public panics because all hydrocarbons on the earth are running out and unable to sustain our way of living. Where is this energy crisis? Do you know where I can find it? Can I drive and go see it? Is there a YouTube video that shows it? Either I'm blind or the mainstream media has an agenda going on here.

Aside from the fear mongering and perpetuating the myth of oil scarcity, the inability of the media to get to the truth of the matter always bothers me. But then again...do media outlets really want to get to the truth about issues? Do they want to report on the infiltration of corruption at every level of government? Do they want to investigate corporations that have massive amounts of wealth, power and connections to the media itself and may own them numerous media subsidiaries? I'm not blaming the majority of newscasters, reporters and journalists but rather the management power structure that decides what makes the cut and what goes in the trash. Take the April 2010 BP oil spill for example—a tragedy in itself with the loss of 11 workers. Coverage and reporting of the accident as a whole is done well, but it's the little things that bother me. One issue in particular is the pressure. What is the pressure of oil escaping at the beginning of the accident? I don't know. Most of the news articles dealing with psi pressure are in June and July talking about 6000 psi. There's a notable gap in April and May of any official pressure readings. And most of the psi reports from these two months are from public blogs, not exactly credible. The problem is the water depth of the area is about 1.3 km deep,[30] which means the pipe itself goes at least 9.4 km into the earth.[31] Any geophysicist, geologist or scientist knows this kind of pressure is impossible for the depth. The pressure has to be at least 2700-4000 bars (39,000-58,000 psi) at any given area under ten kilometers of the earth.[32] The blowout preventer (BOP) on the rig is Cameron's annular BOP with a psi strength of 2000-20,000 psi,[33] not even close to the real pressure coming up from the earth. Mainstream media outlets fail to report the blowout prevention psi strength and the real psi of the upwelling oil. One possible reason is that it may dislodge the myth of fossil fuels and show the abundance of oil underneath every ocean. Oil corporations know this and purposely tread lightly in extracting oil below the ocean; not because of environmental laws, but rather...there's too much oil. Oil corporations have to report these amounts in their annual reports and pay tax on it. They know that they don't need to find more oil; they need to get the price of a barrel of oil at a much higher level so they'll make more profit from each barrel. Increasing the price when the Earth's supply of oil is so vast demands justification, and this is why the myth of peak oil and scarcity are necessary. A company doesn't need to bring up a barrel of oil from the ocean they don't need for $100.00 barrel when they can do it 10 years later at $500.00 a barrel. The oil isn't going anywhere, and since the oil corporations already own the mineral rights they have nothing to worry about. The mainstream media's inability to make these simple connections is what bothers me about mainstream media.

This is one reason why I turn to alternative media. It's not that mainstream media, or as conspiracy theorists like to say, the "corporate media," is corrupt. It's that alternative media presents information and knowledge in a different framework. But why do mainstream media organizations shy away from making the conspiratorial claims? It's not because of fear. Mainstream media has no problem engaging in controversial topics. But to speak of the abiotic oil theory as controversial is incorrect. Controversial issues are ones that go up against the status quo doctrine, with active citizens, local groups, NGO's and various organizations defending their principles. A few examples are abortion, stem cell research, same-sex marriages, 9/11, child soldiers, assisted suicide, animal experimentation and cloning. The abiotic oil theory is not yet able to fit into the category of controversial issues because it doesn't have a group of individuals actively defending it. It's more accurate to say the abiotic oil theory doesn't meet the threshold level of controversial issues. Aside from five books advocating the theory, individuals that say oil will never run out are vocal in their claims but quick to hide behind the shadow of the fossil fuel theory. We shouldn't judge too harshly though; nor does it make sense to condemn such individuals for not taking the necessary leap into the abiotic theory, which is well below the controversial threshold. These authors and speakers have rent, mortgages, and most importantly, families to consider before making such wild claims in the unforgiving public arena.

The Hubbert Curve

While the fossil fuel definition has a foundation in the minds of students from grades 5 all the way up to university, it doesn't exactly state we're going to run out of fossil fuels. But the obvious implication in accepting this definition is that we have to run out of oil because there's only a finite amount of decomposing dinosaur flesh and vegetation on the planet. And this is where the Hubbert Curve comes in for the kill. I can describe the Hubbert curve and peak oil theory to you in a very simple manner. Visualize a mountain, any mountain you've seen in person or in a picture. Look at the bottom left of that mountain, note the rough and smooth curves of the rock. And then look towards the top of the mountain, as it stands up with pride into the clouds. And finally look at the bottom right, and note the final smooth and rough curvature. Congratulations, you now understand the complex economic formula of peak oil theory.

Marion King Hubbert is the father of the peak oil theory. It all starts with him and will end with this book. Hubbert creates a theory that oil production is like an upward slope where production rises and peaks at a particular point and then falls until there's no supply left. The entire Western economy is operating on this concept in theory (in Chapter 4 you'll see this isn't really the case). Hubbert says oil peak's in 1965-1970 for the US and for world oil production in 2000.[34] Unfortunately I don't have reliable statistics to validate what the production of oil is from 1960-1990. The statistics do exist and various journals and newspapers cite them, but the question of credibility is what prevents me from relying on them. So I'll have to start with 1980. In 1980 the production of oil for the US is 10.8 million barrels per day; in 1999 it hits a low of 8.9 million, and in 2011 stands at 10.1 million.[35] So there's a high, a low and another high similar to 1980, pretty much maintaining a steady level. How does Hubbert's theory account for this? It can't. According to the theory it has to drop. And this isn't nitpicking a year after 1965 and intentionally looking for errors, this is 32 years of data. But this is only for the US. For world oil production the peak date is 2000. From 1980-2011 world oil production is only increasing. In 1980 production is 63.9 million barrels a day and 87.1 million in 2011.[36] There's no gradual decrease, no decrease at all, and no period where oil production is holding steady. It's only increasing. So how does Hubbert's theory account for this? Again, it can't. The only retort future generations of peak oil theorists can say is that OPEC countries (Middle Eastern and South American) are screwing up the curve. This is a desperate attempt to hold onto a theory that a child can draw on a piece of paper and call it a masterpiece. Parents will hang this drawing up and praise the child for their efforts. It is a masterpiece, but only to the child and parents. To the outside world it's worthless.

Am I being too harsh and unfair against the peak oil theory? Perhaps, so I'll back it up with another organization. The Cambridge Energy Research Associates (CERA) is a US consulting firm that specializes in energy markets such as oil, coal, natural gas, nuclear, solar, water and any other resources that can provide energy to an economy. Shocking the world in 2006 their publication, *Why the 'Peak Oil' Theory Falls Down*, goes against the grain:

> *Despite his valuable contribution, M. King Hubbert's methodology falls down because it does not consider likely resource growth, application of new technology, basic commercial factors, or the impact of geopolitics on production. His approach does not work in all cases—including on the United States itself—and cannot reliably model a global production outlook. Put more simply, the case for the imminent peak is flawed. As it is,*

production in 2005 in the Lower 48 in the United States was 66 percent higher than Hubbert projected.[37]

CERA doesn't advocate the abiotic theory. However, I believe they want to claim oil will never run out, but that's just not good for business:

> *World oil production will not peak before 2030—and the idea of a peak, at least as commonly presented, is itself highly questionable. In all three scenarios there is no peak in world oil production (including unconventional oil) caused by a lack of belowground resources. Indeed, the concept of a peak in oil production is misleading.*[38]

Authors who advocate the peak oil theory are less vocal today about their claims of peaks and running out of oil, but nonetheless stand firm to the theory. Why? Simple. When authors write numerous books, hold conferences and make a living by promoting the peak oil theory, if they decide it doesn't work anymore and throw it in the trash, they're also throwing their career in the trash. Such authors don't have a malevolent intent to advocate peak oil, but rather, their families and financial security depend on it. To do away with it puts their family's future at risk. This means that they can never let go of it.

Presenting the evidence and contradictions in Hubbert's theory is sufficient enough to disregard it as nonsense and throw it in the garbage. But there's one more reason to doubt the theory and it's more important than all the prior evidence. Like all damning evidence it's simple and makes perfect sense to the average individual. Marion King Hubbert is an employee of the Shell oil corporation from 1943-1964. He creates his theory of peak oil in 1949 and 1956 while working for an oil corporation. It's like a vacuum salesman telling you vacuums are the best and to stick with carpet instead of hardwood or tiles. Working for the vacuum company, the salesman has to tell you this to get the sale and cash in more profit. By working for an oil corporation Hubbert has a vested financial, social and career interest in creating the theory of peak oil. Any information or "knowledge" about a theory of oil from a research employee is suspicious, especially if it allows the employer to benefit. And the oil corporations definitely benefit from the peak oil theory. By promoting the myth of scarcity, the oil industry keeps the prices high and profits flowing regularly.

For those who still can't do away with Hubbert's peak oil theory visualize this hypothetical scenario, a doctor tells newborn parents their baby has a pre-determined finite amount of breaths once it's born. This is done via a new technological device. And once it runs out no matter the age or health, the person dies. The parents argue that their child will prove the doctor wrong and live past that moment. The doctor says "the technology can't be wrong." This is what peak oil theory is about, a pre-determined amount of time to use oil and then when it runs out it's over for good. In the end, common sense is the real solution to put an end to this Malthusian concept of scarcity. But in the meantime people want and demand evidence in order to prove the fossil fuel theory is incorrect, and that hydrocarbons like oil, natural gas and coal are abiotic. And that's exactly what I'll show you in the next chapter.

Chapter 3:

The Evidence for Abiotic Oil & Against the Biogenic Process

There are no real paradoxes in science; the apparent paradoxes are merely nature's polite way, sotto voce, of informing us that our understanding is incomplete or erroneous.

Thomas Gold, The Deep Hot Biosphere: The Myth of Fossil Fuels.

While the fossil fuel theory comes into existence in 1759 it takes a hundred years for the abiotic theory to come out of the shadows. The courageous genius is none other than Dmitri Mendeleev. *Never heard of him.* Sure you have, from your high school chemistry class. Mendeleev is the creator of the periodic table of elements, the basis for all scientific inquiry. It's unimaginable to think of where our modern world will be without an understanding of the elements. Due to his extensive work in the petroleum industry Mendeleev is one of the most credible sources for the abiotic theory in his 1877 publication, *L'Origine du Petrole* (The Origin of Petroleum):

> *So what are these hydrocarbons? It is probably that these are precisely those of oil. White iron treated by acid gives, it is true, other carbides [carbon compounds] less rich in hydrogen, but if these carbides formed under the influence of high temperature and high pressure conditions (carried out in the depths of globe), they would not fail, according to Mr. Berthelot to become saturated hydrocarbons, similar to those of oil. Thus, starting from the fact that the oil cannot have an organic origin I sought to explain otherwise its production mode. The explanation I was given from the disposal of oil sources, the proportion of metals was likely inside the earth, by the passage of water through cracks and by the action of metal carbides on water, such as cast iron.*[1]

This is a translation using Google Translate. My apologies if the translation is not perfect. However, if you can read French feel free to read the entire publication.

In the 20th century the brilliant Russian Geologist, Nikolai Alexandrovich Kudryavtsev, rekindles the abiotic argument. Unfortunately all of his work is in Russian. As strong as his efforts are in advocating the abiotic theory, Western society tunes him out without the general public hearing a whisper of his evidence. But thanks to Kudryavtsev the stage is set for the American heavyweight astrophysicist, Professor Thomas Gold. From the late 70's until his death in 2004 Gold is the strongest scientific advocate of the abiotic theory and crucial to the foundation of this book. Gold sets the standard for raising hell as a brilliant astrophysicist who constantly interferes with other disciplines and their foundational theories. One in particular is in the late 40's when he sets out to prove auditory physiologists are wrong about how the human ear discriminates sound. The scientific community shuns and ignores him for 40 years until a new generation proves his theory correct, that the ear itself discriminates sound instead of neural processes in the brain.[2]

But that's just an itch for the professor. The real confrontation he wants is to go up against the impenetrable petroleum scientific industry and advocate the abiotic theory of hydrocarbons. And you'll see his efforts and dedication in the following evidence I present for abiotic oil.

Biomass

Biomass is any type of living or recently living organism in use as a source of energy such as firewood, compost, industrial waste, and biofuels. Biomass can produce electricity in combination with coal, or as a biogas, to substitute for natural gas. Most biomass commercial applications are for biofuels such as ethanol, biodiesel and biogasoline. Henry Ford's 1908 Model–T car supports both biofuels and conventional crude oil.[3] The reason why biomass is important to the abiotic theory is not because it promotes abiotic oil, it's because it goes against the biogenic theory of creating petroleum oil. Making biofuels takes a matter of days through distillation and filtration, not millions of years. The error in understanding both oil from the earth and oil from biofuels is not due to common sense, but an incorrect ideology that students and the general public believe. The correct theory is that hydrocarbons form billions of years ago in the formation of the Earth and continue to this day. There isn't a day that goes by that the Earth isn't producing hydrocarbons. The oil formation process doesn't take millions of years, but instead, it starts out billions of years ago. The misunderstanding between millions of years of processing vs. the initial start billions of years ago is why the false ideology exists today. Common sense, without the manipulative ideologies, will lead to an understanding of the natural formation of hydrocarbons in the earth alongside creating biofuels in the lab. If you experiment with making biofuels at home and have the product in a few days, what rational reason is there to say the earth uses a slower process of millions of years? There is none. It's only due to deceptive ideologies by the schools, government and mass media. It should be clear that creating petroleum oil, as in oil from the earth, can't be done in the lab using biological means. Yes, ethanol, biogasoline and biodiesel exist, but the chemical makeup is slightly different. The key is that all biofuels have oxygen in their makeup while hydrocarbons are combinations of only hydrogen and carbon. If the biogenic process of petroleum oil is correct, it should form in the lab using only biological material. The fact that it can't be done doesn't prove the abiotic theory of oil is correct, but it does prove the biogenic explanation of how oil forms is incorrect.

Oil Wells Refilling – Kudryavtsev's Rule

Oil wells refilling themselves are the most compelling evidence to the average individual. It doesn't happen right before our eyes, but it happens nonetheless. There are three reasons why no one is hearing stories about oil wells refilling: one, doing so supports the abiotic oil theory in opposition to the myth of scarcity (it's not profitable to do so); two, oil corporations keep producing oil on site for decades without moving, so people can't go and see the wells refilling because the wells are never empty; and three, the current economics of the oil market don't allow drilling deeper to be profitable. The general public can visit previous oil wells in the US, but they're now shut tight with seals. What they can't see is that a few kilometers deeper the oil is sitting right there constantly refilling itself slowly. Professor Adry Bissada, a former geochemist for Texaco, now a professor for the University of Houston's Earth and Atmospheric Sciences department, explains in a 1995 *New York Times* article:

> *'I think we pump oil out much faster than oil can come in,' he said. 'But from a long-term perspective, I believe that hydrocarbons are coming in from great depths and are filling the newer reservoirs at shallower depths. It's impossible to put a number on the rate at which this goes on,' Dr. Bissada said, 'but I could imagine that this kind of stacked reservoir system, with favorable geologic plumbing between the reservoirs, might refill the upper reservoirs in, say, 10 or 20 years. If we were to go back to some old oil field that had been abandoned 50 years ago, we might drill a test well, and we might find fresh oil. The trouble is that that kind of experiment is too expensive in the present economic climate.'* [4]

Kudryavtsev, a Russian geologist, advocates for the abiotic theory so heavily that Kudryavtsev's Rule is the result of his efforts. The rule states that any area that has hydrocarbons present will have greater deposits the deeper one drills.[5] He cites numerous examples in the Middle East, Vietnam, Venezuela and in Wyoming US (unfortunately all in Russian publications). In regards to Vietnam, the Bach Ho or White Tiger Oil Field strikes oil in 1975 and is still producing oil. A 2003 CNN article puts the estimate at 600 million barrels of future production.[6] In 1930 Wyoming's Lost Soldier Tensleep Oil Field begins production,[7] 80 years later it's still producing oil. This is an interesting example because there are breaks in the timeline where the oil stops and after a period of lag begins to refill. In the 90's injecting CO_2 is done to increase the output, and it works, pushing the amount from 2,500 barrels per day (bpd) to 10,000 in a year.[8] It's currently producing 5,850 bpd and 1,250 bpd of natural gas liquids.[9] Not enough? Let's go to the Middle East. Everyone knows the Middle East is overflowing with oil on the surface, and it's unimaginable how vast the deposits are at depths of a few kilometers. In the April 1995 article of *Offshore Magazine: Why the Middle East Fields May Produce Oil Forever*, the authors show that oil is still going strong in great quantities despite oil production beginning in 1908:

> *As much as 10-20 million bbl of oil have been pumped out daily for over 30 years, and yet the oil reserves are increasing. After investigating the logging analytical data of the lower Cretaceous marine shale, and Jurassic fine-grained marine carbonates below, drilled along the eastern Mediterranean coast and over the subduction zone, Bein and Sofer (1987) found that the marine shale could not be the source of the oil discovered along the coast of Israel. They also concluded that although there is some similarity between the chemical compounds found in the oil, and those found in the organic materials in the carbonates, the source of oil is deeper.*[10]

The simple fact that oil wells have been in production for 50-100 years demonstrate that oil wells are not only refilling themselves, but the deeper the drill the greater the oil pressure. This means that oil production in one area never needs to stop.

Hydrocarbons in Space

Saturn's moon, Titan, is rich with hydrocarbons such as methane (natural gas) on the surface. But how can this be when the biogenic process clearly states that hydrocarbons are a product of biological matter? Simple, the biogenic explanation is incorrect. An oversimplification without evidence of course, but the evidence exists in common sense. No biological life exists on the surface, nor can it with the temperature dropping to -179°C (-290°F).[11] While microbial and bacterial life may be present on Titan, life generally doesn't operate in such sub-freezing conditions. The surface of our moon in the shade is about the same. But this is not just a bit of methane in the atmosphere like on Mars; this is vast according to the European Space Agency (ESA):

> *Saturn's orange moon Titan has hundreds of times more liquid hydrocarbons than all known oil and natural gas reserves on the earth, according to new Cassini data. The hydrocarbons rain from the sky, collecting in vast deposits that form lakes and dunes...The dark dunes that run along the equator contain a volume of organics several hundred times larger than Earth's coal reserves.*[12]

So how can it exist when there's no biological life enabling photosynthesis from solar energy? Because it naturally forms in the planet and is accumulating more every day from its initial formation. The same is true for Earth; no biological material is necessary to create it. On Earth, biological material attaches itself to the hydrocarbons as it rises to the surface. One common criticism to this theory is that hydrocarbons in space carry more carbon-13 (C-13) than the usual Earth carbon-12 (C-12).

Isotopic fractionation is complex and a bit too technical for this book, but I'll explain it in a simple manner. Carbon can exist in 16 different isotopic forms, but only three are naturally occurring (C-12, C-13 and C-14). The carbon is the same, the number of protons (atomic number) is the same for all, but the difference lies in the number of neutrons in the nucleus. For C-12 the ratio is 6 protons and 6 neutrons; in C-13 it's 6 protons and 7 neutrons; and in C-14 it's 6 protons and 8 neutrons. The Earth has more C-12 than C-13. Professor Gold explains how:

> *In earth materials, about one in a hundred carbon atoms is the heavy isotope, C-13; the rest are C-12. No chemical action can change one of these stable isotopes into the other. The ratio than is seen is inherited from the nuclear processes in stars that assembled this carbon.*[13]

The key in answering this criticism lies in the proximity to the sun. The Earth is closer to the sun than Saturn's moon, which means it'll produce carbon naturally in the form of C-12. Titan is farther away from the sun so it will naturally produce C-13 (carbon conducive to the cold environment of the planet). Both C-12 and C-13 naturally form itself in the respective planets. The isotopic fractionation of one or the other is not due to life being present on the planets, but rather its proximity to the sun.

Furthermore, this fractionation occurs in a state of passive diffusion,[14] not by the photosynthetic life process that humans and all living creatures at the surface are familiar with. The key reason why C-13 is less prevalent on the Earth, aside from the natural formation of C-12 in proximity to the sun, is due to its mass. C-13 is heavier than C-12. The significance of this implication is the final piece of the isotopic puzzle:

> *Methane molecules bearing the heavier isotope of carbon would ascend more slowly through the rock than would molecules bearing the lighter isotope. Diffusion from pore to pore would be slowed. The longer transit time would subject those heavier molecules to more opportunities of oxidation. The carbon component would thus oxidize into carbon dioxide, which would continue the ascent.*[15]

Both C-12 and C-13 play a role in carbon gases such as methane (CH_4) and CO_2. Surface dwellers like humans, animals and plants prefer C-12, while the Earth's atmosphere prefers C-13 because of the slower geological process of C-13 occurring over billions of years, predating human footprints. The hydrocarbons in planets like Titan and others in our solar system make sense because of its formation in proximity to the sun and the geological isotopic fractionation process. Natural gas on the Earth and every other planet is abiotic whether it's C-12 or C-13.

Transforming Coal into Oil

The transformation from coal or natural gas to crude oil is an ingenious scientific process that deserves significant recognition. Franz Fischer and Hans Tropsch (German scientists), develop the Fischer–Tropsch process. In February 1930 they receive a patent for their hard work, and in WWII the process moves into commercial production.[16] This is the usual explanation most authors present in books. However, these two scientists aren't the first to realize this. The chemical process in transforming coal to crude oil first begins in 1853:

> *Merrill had been visiting a number of railroads, cotton and wool mills, and oil distributers throughout New England trying to introduce a new lubricant called Coup Oil. Manufactured from the distillation of coal tar, the residuum left in the retorts after distilling coal gas, Coup oil had been patented by Luther Atwood in March 1853.*[17]

It's possible that the Fischer-Tropsch process isn't in use in 1853, but historians definitely can't make the claim that transforming coal into crude oil begins in Germany just prior to WWII. The above quotation is from a 2008 book, quite a long time from the 1850's. The credibility of the quotation is a possible issue for historians. Luckily, Ida M. Tarbell, a journalist, backs the claim up in her 1904 book:

> *Luckily not all the men who undertook the manufacturing of petroleum in these first days were inexperienced. The chemists to whom are due chiefly the processes now used Atwood, Gessner, and Merrill had for years been busy making oils from coal...Merrill at the time was connected with Samuel Downer, of Boston, in manufacturing oil from Trinidad pitch and from coal bought in Newfoundland. The year oil was discovered Mr. Downer distilled 7,500 tons of this coal, clearing on it at least $100,000.*[18]

While the true origin in transforming coal into oil is from the 1850's some authors argue a more conspiratorial tone in suppressing it. While this may be true my research doesn't show this to be the case. In 1952 the South African energy and mining corporation Sasol, begins using the Fischer-Tropsch process, which continues in the 70's, 80's and even today.[19] There are also many US Department of Energy (DOE) reports on the process,[20-22] so I'm unable to justify a conspiratorial intention to withhold information or shroud it in the veil of secrecy.

Though secrecy is not an issue, the fact that a non-living substance like coal or natural gas transforms into crude oil in a matter of days shows the million year biogenic process to be incorrect. While this is evidence against the biogenic theory, it also supports the abiotic process because crude oil doesn't form using any biological matter, it's a chemical reaction that labs can reproduce in days, not millions of years. If you take a piece of coal and transform it into the same hydrocarbon structure of crude oil in a few days how can you say the result is a fossil fuel? *Because coal is a fossil fuel.* Unfortunately this isn't the case either. Coal forms via the abiotic or a combination of abiotic and biogenic processes. See Chapter 8 for clarification.

Sweden's Drill

Professor Gold's crowning achievement in proving the abiotic theory correct costs $25 million and is done in the Siljan Ring area in Sweden. Direct observational evidence is what scientific critics demand and Gold doesn't disappoint. After years of sales pitches Gold gets the okay to drill into an igneous rock area, a space where no sedimentary biological materials can exist. It's just pure hard rock. Reaching a depth of 6.7 km Gold's exploratory well finds hydrocarbons, helium and pentane.[23] The drill stops at this depth due to a magnetite paste blocking the drill from going further. While no commercial well begins as a result, the findings are conclusive—that hydrocarbons are abiotic. Gold tries to get more geologists to look at the results but only receives the silence of one hand clapping:

Our invitations to major petroleum research organizations to send delegates to the site and observe our actions went unheeded. Eighty-four barrels of oil are meaningful, especially when they are found in a location where, in the conventional view, not a single drop of oil could have a rational explanation.[24]

It takes Professor Gold a significant amount of time, resources and finances to get twelve tons of petroleum oil from solid rock; not exactly economically feasible. But that doesn't change the truth. It costs $25 million to prove without a doubt that oil is abiotic and is under igneous rocks with no sedimentary material whatsoever. This is why every textbook that teaches students from grades 5 to university about oil coming from vegetation and animal remains has to change. Out of respect for Professor Gold universities should at least make a foundation grant of $25 million to fund new research to understand the abiotic process.

Chapter 4:

Are We Running Out of Oil?

Those [hydrocarbons] are not running out. In fact, we have hardly started to develop them.

Clive Mather, Former CEO of Shell Canada, 20/20 Interview, May 2006.

The greatest direct evidence for the abiotic theory without the need for scientific investigation is common sense, or more accurately, the abundance of hydrocarbons. There's so much hydrocarbons in the earth that it's impossible for humans to use it out. Does the general public think hydrocarbons like oil, coal and natural gas are running out? The mass media machine and the education system tells us we're running out, but common sense overrides it. No one goes to work or comes home worrying about whether the gas station will have enough for their car. When buying a car at the dealership a buyer doesn't think they'll only get to drive that car for five years and then the good times are gone. The action itself in purchasing a car requires an individual to believe oil will never run out. If oil is scarce the general public will think twice about that purchase and car sales will come to a grinding halt; but of course none of this is happening. Even peak oil theorists don't believe oil is running out when they drive their cars to give speeches on television or in universities. I dare anyone to find a peak oil theorist who is unsure whether they'll have enough gas to fill their vehicles for a decade. Car dealerships and automakers don't believe oil is running out and a scarce commodity; if they do there's no need to design the new 2013 models or the prototype 2014 models. The action of constantly designing new energy efficient vehicles that still run on gas proves that the multi-billion dollar automaker industry doesn't believe in the scarcity of oil. They may not come out and say it, but their normal day to day actions speak louder than words. Passengers on board aircrafts taking transatlantic flights to other continents don't believe in the scarcity of oil. They don't worry about the plane not having enough fuel for their 8 hour flight. The aircraft and aerospace industry also doesn't believe oil is running out. Aircraft manufacturing is a billion dollar industry and if they believe oil won't last for another few decades they'll definitely scale back operations. But instead or pulling back, the orders are rolling in. Boeing's new 737 max jets have over 600 orders and each one costs about 100 million US dollars, that's $60 billion alone!!![1] Virgin Atlantic Airways is small in comparison to Air Canada or Southwest Airlines yet it's ordering $400 million A380 double decker planes.[2] This is a clear message that no one in the airline manufacturing industry worries about the scarcity of oil.

The Canadian, American and British air force don't believe oil is running out any time soon. The defense departments for all three countries are waiting for the delivery of the F35's.[3] These jets will replace older models, and none of them have solar or wind energy powering them, it's pure jet fuel (hydrocarbons). On the surface, governments have to maintain the illusion of scarcity in cooperation with the oil corporations; but the actions of various branches of government demonstrate oil is not going anywhere and is the lifeblood of Western society's infrastructure.

JET REFUELING STATION
JET FUEL
Hey Jake, do we have enough for the big one?
PLEASE USE JET FUEL ASAP!!! RUNNING OUT OF STORAGE!!!
Jet fuel
Jet fuel
Jet fuel
Emergency overflow valve
Jet Fuel
Jet Fuel
Jet Fuel

The Abundance of Hydrocarbons

While this book's focus is on oil rather than other hydrocarbons, it's helpful to go into the abundance of natural gas and coal as well. I'll start with natural gas.

Measurements of natural gas are normally in per cubic feet in the US. Canada and the rest of the world measure in cubic meters and gigajoules. One gigajoule is equal to 1000 cubic feet. As a Canadian I don't understand why Americans measure natural gas in cubic feet. Measuring such vast quantities in terms of feet is like measuring the state of Texas with inches. The results will be monumental calculations going past the trillions. There are thee categories to quantify natural gas: production (commercial), reserves, and undiscovered. The reason why undiscovered is a classification is because the natural gas is so deep in the earth the current technology is unable to bring it up in an economically feasible manner. Most estimates take into account production on a regular basis and the reserves that are recoverable. The U.S. Geological Survey (USGS) claims there's 1,429 trillion cubic feet of natural gas reserves in the world, not including the US.[4] What's the correct pronunciation? The next unit after trillions is quadrillions—1 quadrillion 429 trillion cubic feet. The global general public doesn't usually go past trillions in counting money, yet measuring natural gas in the small unit of feet necessitates its use. The mainstream media usually pronounce it as one thousand trillion, but doing so creates confusion for some and inaccuracies in general. Whenever the public sees such numbers it's more valid to say the figure is over 999 trillion cubic feet. Is 1,429 trillion a lot? It sounds like a large number, but really...it's not even close to the actual amount of natural gas that exists.

Natural gas is the most abundant hydrocarbon in the earth. It's coming up out of every pocket and crest in the earth naturally, especially in the Arctic where no one knows what exactly to do with it.[5] Natural gas (99% methane) is the cleanest of the big three hydrocarbons but doesn't have the luxury commercial value that oil enjoys since 1850. It's only in the past three decades that energy speculators see natural gas as a commodity having tremendous potential to not only heat homes but can work for everyday applications. The infrastructure to put natural gas on par with oil doesn't yet exist, mainly because oil is easier and more convenient to utilize. Oil can exist as more than 10 different fuels, while natural gas has to stay as a gas. It can be liquefied artificially by reducing the temperature for overseas shipments but it's quite costly. Natural gas corporations also retrieve natural gas liquids (NGL's) such as propane, pentane, butane and hexane as by-products; but again, this is only in use on a small scale. The vast amount of natural gas is not in use to any significant degree with the current technology and economic situation. I'll use statistics from the EIA to demonstrate the abundance of natural gas, and that the amount the USGS claims isn't even close to the true quantities of natural gas.

World Natural Gas Production (in cubic feet):[6]

Year	Quantity
1980	53,374,000,000,000
1981	54,737,000,000,000
1982	54,663,000,000,000
1983	55,144,000,000,000
1984	60,027,000,000,000
1985	62,297,000,000,000
1986	63,578,000,000,000
1987	66,584,000,000,000
1988	69,846,000,000,000
1989	72,157,000,000,000
1990	73,787,000,000,000
1991	75,013,000,000,000
1992	75,108,000,000,000
1993	76,505,000,000,000
1994	77,068,000,000,000

1995	78,104,000,000,000	
1996	80,478,000,000,000	
1997	80,428,000,000,000	
1998	81,874,000,000,000	
1999	83,850,000,000,000	
2000	87,099,000,000,000	
2001	88,844,000,000,000	
2002	90,561,000,000,000	
2003	93,423,000,000,000	
2004	95,637,000,000,000	
2005	98,393,000,000,000	
2006	101,993,000,000,000	
2007	104,161,000,000,000	
2008	107,922,000,000,000	
2009	104,830,000,000,000	
2010	111,844,000,000,000	
Total	2,479,329,000,000,000	(31 years)

The world production of natural gas from 1980 to 2010 is 2,479 trillion cubic feet; or more accurately, 2 quadrillion 479 trillion cubic feet. And this is natural gas from the production category, (not using reserves or undiscovered amounts). It's clear that the USGS world estimate (not including the US), of 1,429 trillion cubic feet is incorrect. To think that the world has a finite endowment of 3,908 trillion cubic feet of natural gas from dinosaur and plant remains is of course nonsense. Natural gas is the heart of all hydrocarbons naturally forming itself in the earth and will last for hundreds of thousands of years.

I'll move onto our oldest friend that's providing the global general public with electricity for the last 250 years, coal. Quantities of coal are in the categories of identifiable, reserves, and speculative resources. Coal is easily identifiable just by digging a few hundred feet and hitting the jackpot. More than half of the reserves are recoverable with current technology and market prices. And the last category, speculative resources, are unrecoverable with current technology, but it still exists. The World Coal Association estimates the world's coal identifiable and reserve base to be 860 gigatons (gt) with a speculative resource of 17,167 gt[7]. It's amazing how the world decides on a large unit of measurement for coal with gigatons but is unable to do so with oil or natural gas. A Sherlock Holmes mystery no doubt. How much is 17,167 gt? One gigaton equals one billion metric tons, which means the original amount is 17 quadrillion 167 trillion tons. But that doesn't really help either. I'll put it into perspective for you. The largest elephant weighs 10,886 kg or 24,000 lbs.[8] That's about 10.9 tons. The heaviest plane in the world is the Russian Antonov 225 Mryia with an empty weight of 175,000 kg or 385,809 lbs.[9] That's exactly 175 metric tons. Comparing our original figures, 100 trillion Russian Antonov planes are necessary to equal the amount of speculative resources for coal.

Records for Worldwide coal production can help readers in understanding the vast quantities in the earth for electricity and steel production.

World Coal Production from 1980-2010 (in short tons):[10]

Year	Quantity
1980	4,181,850,000
1981	4,224,420,000
1982	4,381,245,000
1983	4,417,821,000
1984	4,667,304,000
1985	4,893,544,000
1986	5,010,526,000
1987	5,120,796,000
1988	5,236,231,000
1989	5,312,916,000
1990	5,346,679,000
1991	5,019,694,000

1992	4,950,621,000	
1993	4,851,845,000	
1994	4,939,742,000	
1995	5,076,581,000	
1996	5,097,586,000	
1997	5,122,574,000	
1998	5,031,835,000	
1999	4,902,432,000	
2000	4,893,712,000	
2001	5,161,821,000	
2002	5,274,946,000	
2003	5,666,719,000	
2004	6,221,397,000	
2005	6,553,851,000	
2006	6,773,564,000	
2007	7,052,922,000	
2008	7,459,615,000	
2009	7,545,745,000	
2010	7,984,900,000	
Total	168,375,434,000	(31 years)

This amount is in short tons but simple arithmetic will give us 152,747,624,315 metric tons. In 31 years the entire world is producing 152 billion tons of coal, or 152 gt. In comparison to our speculative resource amount of 17,167 gt it equals less than 1% of the total amount. But this coal no longer exists in the present. Humans use it up in the past. The 17,167 amount relates to how much exists in the present. So the 1%, 152 gt, doesn't take any amount away from the coal reserves or the speculative resource amount. In short, we're not running out of anything. The abundance of coal is second to that of natural gas. It has a lifespan of hundreds of thousands of years.

And finally, the last commodity in abundance is oil. Oil refineries can make various fuels such as gasoline, kerosene (jet fuel), and distillate fuel (heating oil, diesel). There are many reserve categories for oil, and that's the problem. So many different categories are the cause of deliberate scarcity myths by politicians, but also unintentional scares. Sometimes public officials have the wrong information and can only act on the information available to them. The following classifications are from the Society of Petroleum Engineers and the World Petroleum Council.[11]

The first classification is production. This type of oil extraction is responsible for oil production in the last 160 years. The second classification are in 3 smaller categories: proved (proven), probable and possible reserves. All three types are recoverable using current technological means and waits patiently to get its turn. The next type is contingent reserves. These are oil reserves that are in the earth but not economical to extract using our current technology. It's recoverable but not profitable enough when the price of a barrel of oil is so low. This classification is more dependent on the market rather than upgrading technological components. A simple example of contingent reserves is when a company pays $500.00 to extract a barrel of oil and then sells it on the market for $100.00, not only negating profit but taking a loss as well. The sixth type is prospective resources. This can be recoverable, but again, just off of commercial production and waiting for future projects to make use of it. The seventh and last type of reserves is unrecoverable. The extraction of these reserves is not possible with current or foreseeable potential technology along with the economic market. Many political statements only take into account oil already in production and proven reserves, not even close to the other reserve classifications that show a massive supply lasting hundreds of thousands of years.

Chapter 2 has the total amount of oil production for the last 32 years. And from that calculation we get 841 billion barrels of oil. The USGS estimates we have 665 billion barrels of oil left, not including the US.[12] The numbers aren't the real problem, it's the different classifications of where this 665 billion is coming from. Are they referring to oil currently in production or proven reserves? Or is it production, reserves and contingent reserves? It's more helpful to the public if the Society of Petroleum Engineers and the World Petroleum Council sets international standards for reporting quantities. Perhaps this book will give them a nudge to do so.

By combining these figures we get 1.5 trillion barrels. Is this Earth's finite endowment? Impossible, since the 841 billion is only for the last 32 years. Oil is pumping commercially for 160 years. I have no idea how much oil exists prior to the 32 years but I'm sure it's over 10 trillion barrels. The truth is that oil isn't going to run out in the foreseeable future. If an individual or a corporation digs deep enough they'll find oil and natural gas, I guarantee it. If new more efficient technological innovations come into existence it'll enable homeowners to extract oil in their backyards. Such a fantasy will change the standard of living for the global general public.

REGULAR SUBURBAN LIVING
AMY's HOUSE
Hey Mike, did you hit oil yet?
MIKE's HOUSE
I hit coal at 1km Amy. How bout you Bob?
BOB'S HOUSE
I hit oil at 15km Mike. Did you get any yet Randy?
RANDY's HOUSE
I hit natural gas at 50km. Come over and celebrate, easy on the matches though.

Look at any natural resource that goes into building our modern infrastructure and you'll find that it's not running out. Builders are using quadrillions of bricks and concrete to build millions of houses and megalithic structures in the past 400 years (not including the previous 6000 years of masonry). With all that use it should be ripe for a shortage, but it isn't. How about copper? Predictions in the 1920's by geologist experts claim that copper will run out in 20 years.[13] 90 years later copper is still going strong. But if it's running out someone needs to run and tell Home Depot to start jacking up the price for their copper pipes from $6.00 to $60.00.

The simple reason why abundance is not helpful to business is just that, a massive supply of hydrocarbons or any commodity creates a lower price in a true democratic capitalistic market. The myth of scarcity has to exist so oil and mining corporations can fully utilize the concept of greed. It's only due to scarcity myths that massive profit comes at the expense of the general public. This is the backbone of Western society's capitalistic democracy. And no one in the past takes advantage of this more than one person on the Earth.

Chapter 5:

The Profit Factor

I have ways of making money that you know nothing of.

J.D. Rockefeller, spoken by Mr. Hewitt in an 1879 Congressional Hearing.

All high school students in grade 12 learn that the problem with the world's resources is not a matter of supply, but rather an unequal distribution of wealth and control. And no one knows more about this than one individual. The man who is responsible for the creation and present state of the billion dollar oil industry, the founder of Standard Oil, is the legendary John David Rockefeller. The single largest oil refinery in America for 31 years is Rockefeller's Standard Oil Company. 31 years may not seem like a long time in the span of oil production since the 1860's, but think of it in modern terms. If McDonald's says no company in the US can make a competing fast food restaurant for three decades the courts will have a backlog of litigation. I'm oversimplifying it a bit too much. No one's saying public citizens can't make a competing oil company in the 1860's, numerous entrepreneurs choose this get rich quick oil scheme. But when a corporation has the market in North America, operations worldwide, the backing of politicians, and a bottomless wallet, it's not really competition anymore. J.D. Rockefeller creates a monopoly that allows competing companies to sit at the big boys table but not allowing them to speak.

Using the term legendary is by no means accidental in the description of J.D. Rockefeller. He's man with no ethics, with a calm demeanor, foresight, and great patience; a dangerous combination for any individual who endeavors to control global resources. Monopolizing oil refineries throughout the US may be a large goal for some, but not Rockefeller, in the 1870's the next target is undoubtedly the world's oil supply (and some argue this already is the case). He achieves his national goals with deplorable tactics first starting in 1862 as a financial backer to an oil refinery.[1] He forms Standard Oil in 1870. The initial scheme in 72' is to merge all the refineries into one, with Rockefeller as the president. Next, Rockefeller monopolizes the transportation of moving oil and receives discount rates, controlling the output of oil to local or foreign markets.[2] The efficiency and success of his first plan is terrifying. Of the 26 competing oil refineries in Cleveland, by 1872 he controls 20 of them because he has the best transportation rates in the business. No one can compete with his special discounts and rebates and make a profit. The competing companies either go bust or Standard Oil absorbs them. Monopolizing all the refineries and transportation rails in Cleveland is a huge success and most men will stop there, but Rockefeller is not most men. Where his insatiable greed to dominate and control comes from is beyond the scope of this book. His actions are definitely not to stabilize the markets for regular prices and fair competition. The price of oil fluctuates then as it does now. In 160 years of commercial crude oil production there is never a stable or regular rate for oil. But why stabilize it when profits come from the fluctuations? In Rockefeller's time he's a multi-millionaire. This is equivalent to a multi-billionaire in modern times. After 1872 his plans to monopolize all other refineries in the US (New York, Pittsburg, Philadelphia, Boston...) move at break neck speed. The creation of front companies swallowing up competing ones and then merging them as one becomes an efficient strategy (unknown to the general public). How does this happen in a free market capitalistic democracy? Secrets. Or in Rockefeller's own words, "Silence is Golden."[3] But in this time period there are still many smaller independent firms just trying to make a living while competing in an unfair market. Rockefeller knows how to deal with them as well. For the small refineries, he lets them go bust, and for the ones that he can use he offers them a deal, but one off the books:

> *According to the testimony of one of the firm given a few years later on the witness stand in Cleveland the contract was signed at night at Mr. Rockefeller's house on Euclid Avenue in Cleveland, where he told the gentlemen that they must not tell even their wives about the new arrangement, that if they made money they must conceal it—they were not to drive fast horses, "put on style," or do anything to let people suspect there were unusual profits in oil refining. That would invite competition. They were told that all accounts were to be kept secret. Fictitious names were to be used in corresponding, and a special box at the post-office was employed for these fictitious characters. In fact, smugglers and house-breakers never surrounded their operations with more mystery.*[4]

While his tactics are beyond disrespectful for another human being to engage in, his ability to keep succeeding bothers me.

Yes, he's a shrewd business man but there's something more. I believe that something...is the compliance of the US government. Despite Rockefeller's unethical practices such as price fixing, discrimination, bullying and intimidation, the government's role to implement fair trade is missing in the 1860's, 70's and 80's. Even more surprising is that it doesn't exist in the modern era. In the early days of oil production, price regulation is non-existent because it's the first time such an economic opportunity exists to make money quickly. It's unfair to judge the industry so harshly in the beginning. But the government can easily step in to regulate the prices and stabilize the market in the 1870's but it chooses not to. Court prosecutions, indictments...none of it sticks, there's always a delay and reluctance for the US congress to pass any piece of legislation that can equal fair trade for the oil industry. Today, the oil industry is merely continuing from Standard Oil's beginning. The fact that it no longer exists as of 1911 makes absolutely no difference—same companies, different forms, same agendas: control and money.

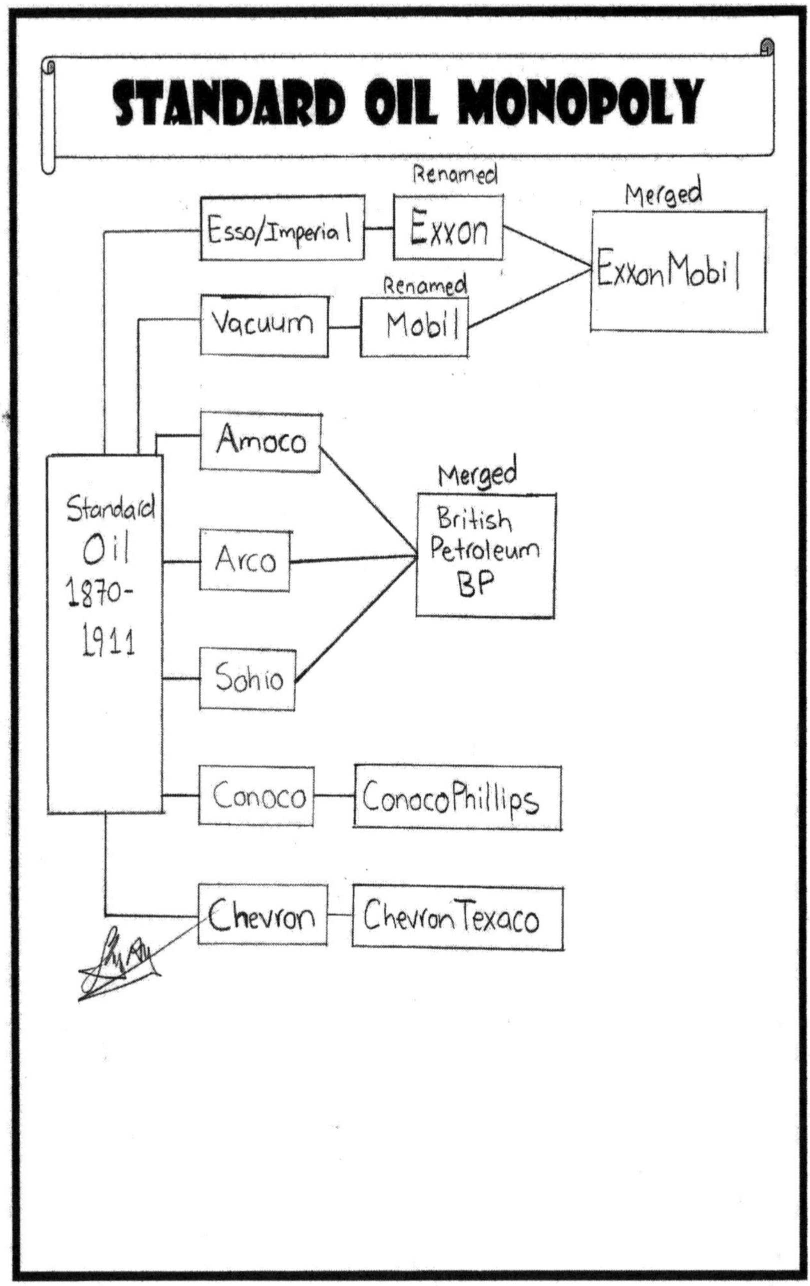
STANDARD OIL MONOPOLY
Standard Oil 1870-1911
Esso/Imperial
Renamed
Exxon
Merged
ExxonMobil
Vacuum
Renamed
Mobil
Amoco
Arco
Sohio
Merged
British Petroleum BP
Conoco
ConocoPhillips
Chevron
ChevronTexaco

Oil Corporations Swimming in Trillions of Dollars

Many peak oil theorists who believe oil is running out and it's "just a matter of time," argue that oil corporations are trying to cash out on the last few barrels. Tell me, where are these last few barrels? Is it coming next year? The next decade perhaps? The simple truth is that if a commodity runs out so does the business and billion dollar profits. Does anyone think oil corporations will let this happen? It's not a good business practice to stop selling a major commodity. But it does make sense to manufacture scarcity and take advantage of natural disasters. Oil corporations make more money when the price of a barrel of oil is high and lose money when it's low. By losing money I mean their profits are slightly below the billions they normally make. This is one reason why there's profit in creating a shortage due to natural disasters or civil war in the Middle East. All of it serves as the excuse to jack up the price of gas at the pumps. It's not that supply is too low, but rather, the supply is too vast and demand is too low to artificially create $500.00 or $1000.00 per barrel. When a refinery is hit by a hurricane oil corporations clink their expensive wine glasses in celebration. Sure it costs them money to repair those refineries, but in comparison to what they'll make by charging more per barrel it's trivial. I'll present the profits of five oil and natural gas corporations below and you can judge whether they have a vested interest in keeping the price of oil high by any means necessary. Remember, this is not revenue per year, it's pure profit.

Year	Royal Dutch Shell[5-15]
1998	$350,000,000
1999	$8,584,000,000
2000	$12,719,000,000
2001	$10,350,000,000
2002	$9,722,000,000
2003	$12,496,000,000
2004	$18,540,000,000

2005	$25,311,000,000
2006	$25,442,000,000
2007	$31,331,000,000
2008	$26,277,000,000
2009	$12,518,000,000
2010	$20,127,000,000
2011	$30,918,000,000
Total	$244,685,000,000(14 years)

14 years of profit equals $244 billion. And remember Royal Dutch Shell has been in business for 105 years. Too much truth? Wait until you see ExxonMobil.

Year	Chevron[16-26]
1998	$1,917,000,000
1999	$3,247,000,000
2000	$7,727,000,000
2001	$3,288,000,000
2002	$1,132,000,000
2003	$7,230,000,000
2004	$13,328,000,000
2005	$14,099,000,000
2006	$17,138,000,000
2007	$18,688,000,000
2008	$23,931,000,000
2009	$10,483,000,000
2010	$19,024,000,000
2011	$26,895,000,000
Total	$168,127,000,000 (14 years)

Year	ConocoPhillips[27-36]
1998	$237,000,000
1999	$609,000,000
2000	$1,862,000,000

2001	$1,661,000,000
2002	Loss of $295,000,000
2003	$4,735,000,000
2004	$8,129,000,000
2005	$13,529,000,000
2006	$15,550,000,000
2007	$11,891,000,000
2008	Loss of $16,998,000,000
2009	$4,414,000,000
2010	$11,358,000,000
2011	$12,436,000,000
Total	$69,118,000,000 (14 years)

Year	British Petroleum (BP)[37-44]
1996	$7,420,000,000
1997	$5,676,000,000
1998	$3,224,000,000
1999	$5,008,000,000
2000	$11,870,000,000
2001	$8,010,000,000
2002	$6,854,000,000
2003	$12,618,000,000
2004	$17,262,000,000
2005	$22,632,000,000
2006	$22,286,000,000
2007	$21,169,000,000
2008	$21,666,000,000
2009	$16,759,000,000
2010	Loss of $3,324,000,000
2011	$26,097,000,000
Total	$205,227,000,000 (16 years)

Year	ExxonMobil[45-53]	
1995	$8,846,000,000	
1996	$10,474,000,000	
1997	$11,732,000,000	
1998	$8,074,000,000	
1999	$7,910,000,000	
2000	$17,720,000,000	
2001	$15,320,000,000	
2002	$11,460,000,000	
2003	$21,510,000,000	
2004	$25,330,000,000	
2005	$36,130,000,000	
2006	$39,500,000,000	
2007	$40,610,000,000	
2008	$45,220,000,000	
2009	$19,280,000,000	
2010	$30,460,000,000	
2011	$41,060,000,000	
Total	$390,636,000,000	(17 years)

A whopping 390 billion dollars of profit over 17 years is enough to give the strongest man pause. Unfortunately I don't have access to data further back. Being a Canadian citizen, I can't file Freedom of Information (FOIA) requests with the US or UK government. Perhaps my American and British counterparts can pick up where I leave off.

Taking into account the former Standard Oil companies (Chevron, ConocoPhillips, BP and ExxonMobil), the amount of total profit over the last 14-17 years is over $833 billion. Not bad for a scarce commodity that's running out? And remember these corporations are in business for the past 50-100 years. 14-17 years of profit doesn't come close to the total trillions of dollars of profit over that span of time. If this isn't bad enough I'll top it off with the retail price of gasoline vs. the average US income.

Looking at historical gasoline retail prices it's mere pennies in comparison to what it is today. But of course that's not taking into account inflation and other financial adjustments. The concept of inflation is deceptive and doesn't provide a clear understanding of prices. Using a different method with nominal dollars will provide a better understanding. Nominal dollars means the actual current value of dollars in use at that time, with no adjustment for inflation.

In 1976 the US retail nominal price of gasoline per gallon is $0.59. In 2011 the nominal retail price is $3.53.[54] In 1976 the US nominal median income is $12,686.00. In 2011 the nominal income is $50,054.00.[55]

I'll use percentages to work out the difference. If 59 cents and 12,686 dollars is our starting block, a 100% increase in both will equal 118 cents and 25,372 dollars. Doing simple arithmetic the results are almost a 600% increase in gasoline prices from 1976 to 2011, and almost a 400% increase in the median income from 1976 to 2011. Gasoline prices are increasing by 6 times whereas income is only increasing by 4 times. If both go up by the same amounts there's no reason to complain. This unequal balance in price and income shows that gasoline is an exception to market democracy.

Gasoline prices are the best example in the failure of the democratic capitalistic process. The laws of the market are supply and demand. Normally, if you have a lot of supply and a lot of demand, the price for a commodity should be relatively stable; not exactly cheap but not too high. If the supply is scarce and demand is high the starting price will be high and gradually increase every year. Gasoline (oil) is the one commodity that breaks all of the rules. It's a commodity that's abundant but operates under the guise of a scarce resource. The demand is always high, so the price should forever gradually increase until oil runs out. But that's not the case. The price fluctuates up and down with every small or large global event. A former dictator falls, immediate price fluctuation. An oil field is set on fire in the Middle East, immediate increase in price. A hurricane destroys an oil refinery, immediate increase in price even though the actual supply of oil doesn't change.

Oil corporation profits are increasing from 300% to 1400% over the past two decades. The price of gas alone is increasing almost 600% to the public with a shortfall of 200% in regular income. If rent is $1000.00 and a person makes $800.00 they won't be able to sustain their lifestyle when the increase in price outpaces an increase in income. It should be crystal clear that oil corporations don't lose. The recession of 2008-2012, Hurricane Katrina, the BP oil spill, continuing civil war in the Middle East...it makes absolutely no difference to oil corporations. The tragedy is that greedy oil corporations are to the left and incompetent governments to the right; either way the public loses. Governments and oil corporations are definitely rigging the game, and not to the general public's benefit. I'm fine with the game being equal on all terms, but the truth is that the current economic system doesn't allow for equal distribution of resources, food or commodities. It's always one or the other that has the advantage, and usually it's the billion dollar corporations that have control.

The strongest indicator for oil corporations having the advantage is to look at a staple of North American lifestyle, the car. Buying a car is a rite of passage for most young adults. They save up their money from their part-time jobs or getting a little help from the parents...all for the goal of purchasing their first car. While this is great for the young adult buyer it's better for the oil corporations. Sure, the car dealership makes money as does the insurance company; but the real money comes from the daily need of the car, gasoline (oil). And it's not just the initial sale that makes the difference, but rather the feeling of loving the convenience a car offers. This is why oil corporations usually have a person for life when they buy their first car. When the car breaks down the consumer will buy another one. Even though car dealerships try to win customer loyalty, it makes no difference for the oil corporations because all cars take the same fuel. There's no way around it; the greed of oil corporations knows no bounds. When thinking about this greed my thoughts go to the 1989 film, *National Lampoons Christmas Vacation*:

> *Remember how I was toying with the notion of suspending the Christmas bonuses?*
>
> *Oh you didn't?*
>
> *Well I...*
>
> *Of all the cheap lousy ways to save a buck.*
>
> *That's pretty low mister, if I had a rubber hose I would beat you until you...*[56]

Chapter 6:

The Myth of Global Warming

Few things annoy me more than to hear people talking about carbon dioxide as being a pollutant. You're made of carbon dioxide. I'm made of carbon dioxide. Carbon dioxide is how living things grow.

Nigel Calder, Former Editor of New Scientist, The Great Global Warming Swindle, 2007.

Is it global warming or climate change? Either way, it starts out in the mid 70's as global warming and morphs into climate change in modern times (though both have their origins in the 70's). It's necessary to use both terms to represent an accurate view of this billion dollar industry with good intentions and bad science. The global warming movement is an incredible force, and when there's dissent or questioning from within, the individual is quickly cast out of the lot through humiliation or defamation. As surprising as this reaction is to some, it shouldn't be. Any type of organization or movement with a connection to money must have stable ideologies. Religious dogma exists, why not environmental?

Global warming has an underlying assumption at its roots. And this assumption carries the weight of all policies it creates, be it political, social, environmental and economic. Without this assumption the movement loses its entire foundation. Simply put, man-made carbon dioxide is causing an increase in the temperature of the Earth.[1] All the billions of dollars, policies and people's hopes rest on this foundation. That's why it's so tragic that it's false.

I have a suggestion for the average individual who is a part of the climate change movement and desires the truth. Go to a climate scientist (as if they're available at your local Wal-Mart) and ask for a graph of the Earth's temperatures and CO_2 levels for the last 500 years (stick with verifiable human history). Ask for a large horizontal printout. What you'll see are curves going up and down. I dare any person to present a graph moving upward in a straight line. When you ask a climatologist why the temperature drops the usual explanation is that not much is happening in human history. Unfortunately this explanation is not valid. An indisputable fact for both sides of the camp is that the temperature of the Earth is rising from the mid 1970's and holding steady in the last decade. No reputable scientist in the world will dispute this. But why am I starting then and not sooner? Because from the 40's to the mid 70's the surface temperature of the planet drops while human activity and industrial production is at its highest during the post WWII economic boom. Japanese professor Syun-Ichi Akasofu, the Director of the International Arctic Research Center, says it more clearly:

> *CO_2 began to increase exponentially in about 1940 but the temperature actually began to decrease 1940...ah continued to about 1975, so this is opposite relation. When the CO_2's increasing rapidly but yet the temperature decreasing, then we*

cannot say that CO_2 and the temperature go together.[2]

In 1975 Dr. Wallace Smith Broecker, the pioneer of the global warming movement, writes an article attempting to explain this cooling as a "natural" climate fluctuation. And this cooling compensates for the increase in CO_2.[3] This is confusing for me. Iron ores, which make up steel, naturally form itself in the earth but oil doesn't? The period from 1940-1975 is a natural climate fluctuation and 1975 onwards is man-made? To say that human activity and industrial production causes the Earth's temperature to rise and glossing over parts where it drops with significant human industrial activity is scientific fraud. Too harsh? Perhaps. But it's at least a great disservice to scientific development and to the average individual who doesn't specialize in this area, but relies on the accuracy of the information. There are thousands of global warming computer models that research teams keep away from prying eyes. The purpose is to hide their formulas from other researchers trying to validate the models. But if you get the chance to see a computer model of the Earth's temperature and CO_2 for the last three decades make sure the data backtracks to the previous three decades. If it can correctly predict the 1940-1975 fluctuation then it's accurate, if not, it's worthless and isn't credible as scientific research.

Al Gore's 2006 documentary *An Inconvenient Truth* shows a long graph correlating CO_2 and the Earth's temperature together. What he fails to mention is the space between the two measurements, a lag. It's not that CO_2 is causing the temperature to increase but rather CO_2 is following temperature increases with a lag in time. Professor Ian Clark explains it at the Canadian Senate Energy Committee hearing in 2011:

And here we're looking at a very detailed interface between a glacial period here at 245 to 240 and an interglacial period. So we have warming shown in red. And that

> *warming occurs about 200...about 800 years prior to the CO_2 increase. So CO_2 is lagging temperature increase by about 800 years. And this has been demonstrated for all these glacial interglacial interfaces through time. So we're always seeing a lag...CO_2 is not driving climate, CO_2 is not acting as a greenhouse gas over this very important period of strong climate change.*[4]

Earth's atmosphere is mostly nitrogen and oxygen, making up 99% of the atmosphere. Carbon dioxide is a greenhouse gas because it's part of a category of other gases that absorb energy that the sun gives off. Is CO_2 the larger of that 1% of greenhouse gases? Nope. That distinction goes to water vapor from the oceans and argon. This means that ocean cycles play a significant role in the temperature of the Earth. Oceans are vast beyond imagination. If you can imagine the corners where the Pacific Ocean meets land and all the water in the middle...that's very impressive. Your visualization skills are far superior then mine. Oceans dwarf humans and all industrial activities by far yet mainstream media attention is minimal. Oceans swaying under the moonlight are a calm and pleasing action, but not exactly a sensational news story. CO_2 makes up about 0.4% of the greenhouse gases.[5] So where is all this CO_2 coming from? According to a 2007 article in *New Scientist*, annual CO_2 comes from the respiration of animals, 220 gigatons (gt); vegetation, another 220 gt; oceans release about 330 gt, and human activity produces 26.4 gt plus 5.9 gt for deforestation.[6] Doing the math, human activity is responsible for 4.2% of the total CO_2 output. But this is 2007 data; I'll be generous and push it to 6.0%, an extra 13.9 gt per year. So are people affecting the atmosphere of the Earth by 6.0% in their daily lives? Nope. Remember, the atmosphere is 99% nitrogen and oxygen, CO_2 is part of the 1% of greenhouse gases. Alright then, so are people affecting the 1% of greenhouse gases? Nope. Carbon dioxide is not the dominant greenhouse gas, that's water vapor. CO_2 accounts for 0.4% of greenhouse gases. One more time. Are people responsible for 6% of 0.4% of carbon dioxide in Earth's atmosphere? Yes. This is why billions of dollars of research, computer models and the entire climate change industry exists...for 6% of 0.4% of CO_2 in the atmosphere. This kind of money can go towards more important scientific developments such as deep ocean research, investigating the nucleus of our galaxy, or understanding the potential of the human mind.

Changes on the Earth, be it temperatures, land masses, floods, volcanoes and hurricanes are all fascinating and capable of causing great damage, but the key point is that they're natural. To say the climate is changing is a natural everyday observational statement. Billions of dollars of climate computer models are unnecessary to say that. It's fascinating and tragic to see how these billions of research dollars beat out general common sense. To say the surface temperature on the planet is changing, is an everyday observation. And to say it's fluctuating and doing so for millions of years costs only single digit millions to look at ice core samples. To determine whether our atmosphere is warming, again, it only takes a few millions in research dollars to create weather balloons that give us accurate information about our atmosphere.

So why do environmental scientists tell the general public their CO_2 breathing and commercial activities make a difference to the Earth's temperature? Exactly that word, that it makes a difference. As a kid growing up in the early 90's I remember watching the cartoon *Captain Planet and the Planeteers,* where five youth have the power of the elements (earth, water, fire, wind and one for heart). They use their powers to help those in need; but when the situation is dire they use their rings to call on Captain Planet (some sort of living incarnation of the Earth in human form), and Captain Planet keeps industrial polluters in check. It's a corny but fun TV show with the message that individuals can make a significant difference to the health and well-being of the planet. Individuals in general like to feel their existence makes a difference while being on the Earth. Telling someone that their breathing and daily activities make it harder on the Earth causes a gasp reaction, *I'm hurting the Earth from my activities?* But it also gives an individual a sense of entitlement and responsibility—that their breathing and industrial activities are so great that it affects the mechanics of a massive planet. This moment of awe is what pushes people to change their actions and reduce their carbon footprint; because it's their carbon footprint that's hurting the Earth, right? Well, 6% of 0.4% anyway.

The real culprit for planetary changes is our shining friend the sun. Why is he/she (humans like to attach anthropomorphic qualities to all things), raising the temperature? I'm not a scientist so I can't give you a scientific answer. I can only make the claim that our sun knows what it's doing; it's been doing it long before humans begin arriving on the planet; and I'm sure it'll keep doing it for millions of years in the future. But if I have to guess, I think it's a matter of solar cycles. Scientists who oppose the man-made theory of climate change usually go with the sun as the single largest factor causing temperature changes in the surface and atmosphere of the Earth. Professor Timothy Patterson presents his findings at the Canadian Energy Senate Committee Hearing in 2011:

> *In that record...ah...we discovered repeated cycles in marine productivity that correlate very well with cycles in the brightness of the sun. And this was not unique and hundreds of other studies have shown exactly the same thing—that the sun and not variations in carbon dioxide...ah the gas that's most targeted*

by Canada's National Climate Change campaigns, appears to be the most important driver of climate change.[7]

While there are other scientists that advocate the solar cycle theory I find that Professor Patterson presents a more objective viewpoint because he's looking for the connection between climate change and fish productivity. Whether climate change happens through man-made or solar cycles makes no difference, the effect on fish is what he's after. He goes on to say:

> *Many things that we thought we knew about climate system just a few years ago are now proving to be highly uncertain or quite mistaken. It's no exaggeration to say that in the period since the Kyoto protocol was introduced there's been a revolution in climate science. If back in the mid 1990's we had known about what we know about climate change today there would have been no Kyoto protocol. Because it would have been considered unnecessary.*[8]

Unfortunately, finding out that solar activity plays a larger role in Earth's temperature causes a deeper question: what determines the various cycles of our sun? Your guess is as good as mine. The only help I can offer is for you to remember that the sun is moving as well. Our sun is orbiting around the nucleus of our Milky Way galaxy. And this may be one avenue where funding can increase. Understanding the celestial phenomenon that affects our sun and in turn affecting the temperature of the Earth is a valid scientific cause.

In researching global warming I find the concept is strikingly familiar to the fossil fuel theory. In global warming CO_2 is the culprit that causes a rise in global temperatures. In the fossil fuel theory microbial life causes the formation of oil. The key to understand both is to switch it and reinterpret. It's not that microbial life creates oil, but rather oil comes up naturally from the earth and picks up microbes along the way. It's not that CO_2

causes a rise in temperature, but that a rise in temperature causes the CO_2 levels to follow.

In a more general sense the concepts are again similar. Hubbert and the succeeding peak oil theorists espouse that oil peaks in the US in 1970 and for the world in 2000.[9] Soon after, we'll run out and won't be able to sustain our way of living from our energy consumption. The same is true for climate change. The April 2006 cover for all *Time* magazines in the world reads "Be Worried. Be Very Worried."[10] The article is from March 2006 (text capitalization is the same as it appears in the magazine):

> *Scientists fear we may be approaching the point of no return THE GREENHOUSE EFFECT Without the greenhouse effect, life on Earth would not be possible....BURNING FORESTS REDUCES OXYGEN AND INCREASES DROUGHT SPREADING THE PAIN Deforestation....Soil dries out RISING TEMPERATURES MELT POLAR ICE AND PERMAFROST THAWING OUT The North Pole may be seasonally ice free by 2050.*[11]

The similarity is in the Malthusian concept of fear and scarcity for both oil and our way of living—that there won't be enough for everyone—that we'll have to revert back to a primitive lifestyle of war, with fear constantly on our minds as we go to sleep. These tactics are both unethical and promote an unhealthy ideology. I know most activists and environmental students are good loving people that want to do the best they can for their families and the Earth. I have no problem with an individual deciding to reduce their carbon footprint—to each their own. But advocating these ideologies to vulnerable populations such as 3rd world developing nations is when I have a problem with them.

These individuals don't have the luxury of civil debates in conference halls with electricity to provide air-conditioning to help cool their tongues. They have the luxury of no electricity, no sofas, no air conditioning, no fridge, but more importantly no development. James Shikwati says it best:

> *The challenge we have when we meet Western environmentalists who say we must engage in use of solar panels and wind energy is how we can have Africa industrialized. Because I don't see how a solar panel is going to power a steel industry. How a solar panel you know...is going to power maybe some railway...train...truck. It might work maybe to power a small transistor right now.*
>
> *One clear thing that emerges from the whole environmental debate is the point that there's somebody keen to kill the African dream. And the African dream is to develop.*[12]

And I believe this "someone" is the Obama Administration, in his own words:

> *Ultimately if you think about all the youth that everyone's mentioned here in Africa. If everybody's raising living standards to the point where everybody's got a car, and everybody's got air conditioning, and everybody's got a big house...ah well, the planet will boil over. Unless we find new ways of producing energy.*[13]

My initial reason for including climate change in this book is to explain the connections between the fossil fuel theory, governments, oil corporations, and climate change from man-made CO_2. I don't know it's coming from billions of dollars of corporate science when I begin my research. It's only from looking at the evidence that my view changes. In my research man-made CO_2 doesn't cause Earth's temperature to rise...okay. If CO_2 does cause Earth's temperature to rise...okay. Either way it doesn't affect the premise of this book in the slightest. This objective viewpoint toward climate change is something environmentalists should take into account before the lynching begins.

Environmentalists, climatologists and the green movement are quick to vilify anyone that doesn't tow the party line and most likely, will say the same about me. If they don't know my beliefs and intentions towards the Earth, their condemnation is justifiable because they don't know me. I believe the planet operates as a living organism that's able to regulate itself without the egotistical protecting arm of humans. Does that mean I advocate an uncaring attitude for people to run amok with the planet? No, the general public, activists, and governments need to hold corporations accountable and responsible for their polluting actions; not because it's killing the Earth, but rather out of respect to the rest of the general public who use it. I support more energy efficient products because it means society is developing better technology. One of my personal goals is to buy a piece of land up North and build a house with my own bare hands. Does that mean I'm against modernity and globalization? No. Getting this far in my life is only possible with the technological means of the industrial revolution. Individuals who choose the city life and the economic benefits it offers such as jobs and security for their families, have nothing to apologize for.

Do I support nuclear energy? Definitely not. If I have to choose, I'll live with all the CO_2 from the coal industry than nuclear radiation in the atmosphere. Is someone paying me to help oil corporations? No. This book is about the abundance of hydrocarbons in the earth and the intentional deception by governments and oil corporations. By advocating the abiotic theory, the myth of scarcity loses and oil corporations' greed can be put in check. I'm pretty sure I won't be making any friends or money with them. And I doubt that I'll generate any sales from oil corporations that "just can't live without this book." The oil industry and the climate change industry are both multi-billion dollar industries. Do you really think they'll just accept defeat because people now desire the truth?

Government & Industry Hijacking Climate Change

The concept of global warming begins in the mid 1970's when the temperature of the earth's surface begins to rise. Why in the 70's and not the 60's? Because global temperatures don't start rising until 1975. Dr. Broecker coins the term *global warming* in 75' and the 1st World Climate Conference is in 1979. But the first real catapult to stardom in the public arena is from the US Senate Hearing on the greenhouse effect and global climate change in 1988. Like clockwork, the United Nations Intergovernmental Panel on Climate Change (IPCC) comes into existence. Global warming research and funding skyrockets after 88' with the mainstream media jumping on the band wagon with its feature articles moving at the speed of light. The second wave of the global warming crisis comes in 2006 with Al Gore's documentary *An Inconvenient Truth*. This is also when the term global warming falls out of favor and the term "climate change" comes into being. This is in response to scientific critics who maintain temperatures are cooling, not getting warmer. By changing the terminology global warming activists reinvent themselves and claim that whether it's going up or down it's because of human activity. This silences some of the past critics because they can't say temperatures aren't changing, they're changing all the time. This leads to a less vocal outcry against climate change and a stronger base for the man-made climate change industry. Journalists in mainstream media have run full steam with sensational stories that put climate change on an apocalyptic level, but why? Why is it necessary to go to this level? Nigel Calder a former editor for *New Scientist* answers:

> *You've got a whole new generation of reporters, environmental journalists. Now if you're an environmental journalist and if....the global warming story goes in the trash can, so does your job. It really is that crude. And the reporting has to get more and more hysterical because there are still fortunately a few hardened news editors around who will say you know this is what you were saying 5 years ago. Ah but now it's much much worse, you know there's going to be 10 feet of sea level rise by next Tuesday or something. They have to keep on getting shriller and shriller and shriller.*[14]

If Ida M. Tarbell's 1904 book about Standard Oil represents the best in journalism, then self-righteous environmental journalists who sensationalize climate change and attack those who are against it...represent the worst.

The reason why I question the foundations and the entire climate change industry isn't because it sounds incorrect or doesn't quite add up. Inconsistencies and bad science happen all the time in scientific research. Individuals die defending incorrect theories because they have a personal, financial or even religious stake in the issue. The initial seed of doubt for me is when the former US vice president becomes the figurehead of climate change and presents it to the general public in 2006. In 2012 US President Barack Obama says, "Climate change is not a hoax."[15] The fact that a US president is saying this means there are too many people still questioning the foundations of climate change. And the final reason is of course the money. Climate change begins as an environmental movement, and transforms itself into a billion dollar climate change industry.

Many authors talk about investments of billions of dollars in the climate change industry; but where's the proof? Where are the receipts? Surely there must be at least one invoice for a few billion, right? Yes there is. The problem is that there are so many research projects under the "umbrella of climate change dollars" that it'll fill an entire chapter to include all of it from 1988 to today. *What a scapegoat tactic.* But I'll give you the general overview and just a few of the current and past research projects.

Funding for the climate change industry is by private, public and international organizations. The largest donor is the United Nations Environment Program (UNEP). Though funding for climate change research begins in the 70's it doesn't take off until 1988, so that's where the numbers will start.

UNEP budget from 1988 to 2011in US dollars[16-24] (1988-2003 graph estimate):

1988	$50,000,000	
1989	$60,000,000	
1990	$75,000,000	
1991	$90,000,000	
1992	$103,000,000	
1993	$115,000,000	
1994	$127,000,000	
1995	$109,000,000	
1996	$110,000,000	
1997	$112,000,000	
1998	$79,000,000	
1999	$80,000,000	
2000	$98,000,000	
2001	$118,000,000	
2002	$119,900,000	*These four are partial
2003	$121,000,000	amounts, the full amount
2004	$59,100,000*	for these years are
2005	$59,200,000*	unknown despite sending
2006	$144,000,000*	requests to the UNEP.

2007	$69,200,000*	
2008	$306,700,000	
2009	$570,500,000	
2010	$446,500,000	
2011	$448,000,000	
Total	$3,669,200,000	(over 23 years)

This $3.6 billion doesn't all go towards climate change since budgets include staff and administration expenses. The point is that any project can reference climate change and apply to any aspect of the environment, so researchers potentially have access to all of this funding.

Many countries also have national climate change budgets for any research with reference to or relating to climate change.

Canada:[25-26]

2007 Clean Air & Climate Change Fund	$1,500,000,000
2010-13 Action on Climate Change	$1,200,000,000

USA:[27-30]

1989 Global Change Research Program	$134,000,000
1990 Global Change Research Program	$659,000,000
1991 Global Change Research Program	$954,000,000
1992 Global Change Research Program	$1,110,000,000
1993 Global Change Research Program	$1,326,000,000
1994 Global Change Research Program	$1,444,000,000
1995 Global Change Research Program	$1,760,000,000
1996 Global Change Research Program	$1,654,000,000
1997 Global Change Research Program	$1,656,000,000
1998 Federal Climate Change Program	$4,000,000,000
1999 Federal Climate Change Program	$4,500,000,000
2000 Federal Climate Change Program	$4,400,000,000
2001 Federal Climate Change Program	$4,400,000,000
2002 Federal Climate Change Program	$4,200,000,000
2003 Federal Climate Change Program	$5,700,000,000
2004 Federal Climate Change Program	$5,800,000,000
2005 Federal Climate Change Program	$5,400,000,000
2006 Federal Climate Change Program	$5,100,000,000

2007 Federal Climate Change Program	$5,800,000,000
2008 Federal Climate Change Program	$6,400,000,000
2009 Federal Climate Change Program	$7,500,000,000
2010 Global Change Research Program	$2,100,000,000
2010 Climate Change Technology Program	$5,500,000,000
2011 Global Change Research Program	$2,600,000,000
2011 Climate Change Technology Program	$5,700,000,000
Total U.S. Climate Change Funding	$89,797,000,000

UK:[31]

2011-15 International Climate Fund	$2,900,000,000

Here are a few specific programs by an organization that has a keen financial interest in promoting climate change, the famous Rockefeller Foundation.[32-36] The amounts are in USD.

1988 Woods Hole Research Center, Massachusetts.
$100,000
Conferences in developing countries on global climate change.

1989 Tata Energy Research Institute, New Delhi, India.
$90,000
Creating an Information and Research Centre on global warming and climate change.

1991 Massachusetts Institute of Technology (MIT), Cambridge.
$54,000
$60,000
Hosting a global warming conference October 3rd, 1991 and expenses for the January 1992 conference "Transition to a Global Climate Change Regime."

1991 Climate Institute, Washington DC. $100,000
Developing country policy-makers in India, Bangladesh, Nigeria and Brazil on the impacts of global climate change.

1994 Center for Clean Air Policy (CCAP), Washington DC. $98,000
Designing a replicable financial structure for the private sector under the provisions of the UN Framework Convention on Climate Change.

2007 Center for Clean Air Policy (CCAP), Washington DC. $900,000
Initiatives for cities to adapt to the negative effects of climate change.

2007 Energy Resources Institute, New Delhi, India. $200,000
For 2008 Delhi Sustainable Development Summit focusing on climate change mitigation.

2008 Stanford University, Stanford, CA. $348,900
Supporting climate change resilience models.

2008 University of Copenhagen. $700,000
Development of high resolution digital maps.

2008 Institute for Social and Environmental Transition, Boulder, Colorado. $570,100
Creating a climate change resilience network.

2008 Brookings Institution, Washington DC. $189,000
2008 Conference: How Will the World's Poor Cope with Climate Change?

2008 Oxfam America, Boston, MA. $507,200
Global policy campaign to support US government policy funding climate change research.

2008 National Religious Partnership for the Environment, Amherst, Massachusetts. $400,000
Supporting US climate change policy.

2008 CNA Corporation, Alexandria, Virginia. $799,700
Supporting climate change resilience policies.

2008 Rockhopper Productions Limited, London, UK. $3,163,500
Creating an 8 part documentary series on climate change resilience.

2008 McKinsey & Company, Inc. New York, NY. $500,000
Studying the effectiveness to adapt to climate change.

2008 Institute for Social and Environmental Transition, Boulder, Colarado. $5,348,800
Supporting dialogue between Asian Cities Climate Change Resilience.

2008 Georgetown University, Washington DC. $500,000
Support climate change adaptation policies.

2009 Christian Coalition, Charlestown, South Carolina. $100,000
Building a Conservative Christian campaign to advance US climate change policy.

2009 Alliance for a Green Revolution in Africa. $440,000
Strategic planning for climate change resilience.

2010 United Nations Economic Commission for Africa. $250,900
Supporting climate change mitigation.

2010 Institute for Social and Environmental Transition, Boulder, Colorado. $1,582,240
Establishing climate change centers in Vietnam.

2010 Federal Emergency Management Agency (FEMA), Washington DC. $1,000,000
Community Resilience Programs.

2012 Mercy Corps, Portland, Oregon. $450,000
Climate change curriculum for teaching resources in Indonesia.

2012 International Institute for Environment & Development, UK. $1,991,160
$2,000,000
Creating the International Centre for Climate Change. Supporting research, publications and multimedia teaching resources.

There are of course many other countries and organizations that fund the climate change industry, but the amounts above (almost $100 billion) give you a rough idea of the backing of this industry. But why? What's the purpose in doing all of this? What's the agenda? It's not as conspiratorial as conspiracy theorists tend to think; but it's also not one of good intentions. I'll explain it clearly. This is my own conjecture, and while I don't have evidence to back it up, it's the best possible explanation. Besides, nobody's dumb enough to have copies of documents lying around that can incriminate themselves and others.

In 1972 the Climate Research Unit (CRU) comes into existence at the University of East Anglia, UK. The United Nations Environmental Program (UNEP) also pops up out of nowhere on December 15, 1972. Sadly, the purpose for both organizations is to legitimize the theory of man-made CO_2 destroying the planet due to fossil fuels. It takes 16 years to produce enough scientific evidence to justify the theory and in 1988 the man-made theory steps up to the world stage for the general public to accept. The general public reacts with disbelief and shock, but the science is solid and with the mass media backing the flames of fear, they buy it. Individuals begin to organize into significant environment green organizations (EGOs); and the general public tacitly empowers them to act in their interests. The EGOs agree with the UNEP and CRU that another specific body is necessary to analyze, understand, and stop the threat of climate change. The Intergovernmental Panel on Climate Change (IPCC) accepts this task. For the first time, climatologists have massive amounts of money with the spotlight shining on them as EGOs trust them with their full confidence. These are not corrupt scientists that have a malicious intent to deceive others; they're good loving people with families like you that are put into an extraordinary situation. If you receive millions of dollars to find answers the general public and EGOs desperately needs, what are you going to do? Give them what they want of course. Continuing on, the IPCC creation is twofold. Yes, it's provides the technical scientific analysis of the climate change issue, but it also unites the EGOs as one. Before this time, environmental activism exists, but only out of reaction to Western government policy in a fragmented sense. When things are good, activists, provide small but constant headaches to various industrial corporations. When things are bad they take things into their own hands and act. Illogical reactions and disorganization are difficult to anticipate and control. Creating the IPCC provides the scientific scotch tape (not strong enough as glue) to hold the EGOs as one; but more importantly, to be predictable...which is much easier to handle.

Uniting a movement together, acting in uniform, is delicate work. While the science provides the justification it doesn't exactly play the role of taking the guilt and anger EGOs feel. The guilt exists for causing harm to the planet, and anger for not knowing about it sooner and taking action. Something else needs to play the role of a villain to keep the movement solid; and that of course is the oil corporations. It's the oil corporations that act as the glue, the binding agent that holds the environmental green movement and the entire house of cards together. Without these corporations playing the role of the villain, the movement can't hold itself together and will fragment into sects, taking matters into their own hands.

Oil corporations have to play a difficult role as well. They can't exactly cater to EGOs but neither can they be too friendly with governments. Governments and oil corporations normally go hand in hand for most issues; they merely play different roles for the general public to pick sides. In this framework picture two governments operating in Western society; one for the public sector and one for the private. Whatever corporation is the strongest (having the most money and the largest reach) gets to manage the private sector of government. This is similar to the analogy in Chapter 5 about Standard Oil being the dominant corporation and letting others sit at the big boys table but not allowing them to speak. Looking at the private industry as a whole, oil corporations beat out all others by far. Even banks are second to them because a bank has paper with the assumption of currency value. Oil corporations actually have commodities of an intrinsic value.

In return for playing the role of the villain for EGOs, a penance is necessary; or more clearly, a tax. But not just any tax, a carbon tax. A carbon tax is a possible solution in the 1988 US congress energy hearing.[37] But why do oil corporations want to put taxes on their own goods? Because under the current economic policy the costs passes onto the consumer at the pumps. And this is where we are now. Remember from Chapter 5 that oil corporations will never lose, be it a recession, an industrial accident or natural disaster, the profits continue to flow in the billions. But it's a difficult impasse this time that prevents things from moving forward. On the one hand government and industry need the EGOs to convince the general public to accept a carbon tax every time they go to the pumps. At the same time the government needs to be the sympathetic shoulder to lean on and ease the general public's financial burden with rebates and income tax breaks. For Western governments to create a national or global carbon tax now, when prices for basic living commodities are high, is not a good idea. It may cause low income individuals to go beyond their breaking point and decide to take things into their own hands—actively trying to reform a corrupt system by any means necessary. This is why a national or global carbon tax won't happen yet. The actions are in sequence but not yet ready for the stage.

But what's the agenda? If a carbon tax isn't the end game then what is? John Travolta's character in the 2001 film *Swordfish* puts it nicely, "misdirection."[38] Think of your world as a large pizza, 12 slices, and the global warming issue is one slice. The more your attention is on that one slice the less you notice what's happening with the other 11 that are rotting away. But even if you see what's happening with the other 11 slices you're less likely to act on them because the damage you're causing the planet with your CO_2 takes priority. What are the other slices you need to be aware of instead of the global warming misdirection? The loss of your constitutional rights is a big one. The loss of your privacy in the digital age is another. The substandard education system that provides the least amount of knowledge to continue the same cog-in-the-machine cycles is a huge slice that you need to be aware of; if not for yourself than for your kids, or your nieces and nephews. One of the most important slices is yourself, your health. The health care system in the Western world is entirely dependent on you being unhealthy. Western doctors have a vested interest in keeping you unhealthy and a patient for life. A multi-billion dollar health care system can't operate if the general public is naturally healthy, that's where prescription drugs and GMO foods come into play. Human rights in the developing world are a tragic issue that never receives the focus it deserves. I don't mean mainstream media attention, NGO's or non-profits, but real decisive actions to go up against a corrupt system that wants to keep billions of individuals in a continual state of poverty and lack of respect.

So what's the solution? How does the global general public get out of this state of misdirection? My solution isn't a pleasant one; it's very difficult and most individuals won't choose it; but it will at least be an honest solution. Of all the players in the climate change industry, the most crucial is environmental green movement. There's no point in having glue if you don't have the initial object that needs sticking. Even climatologists and scientific researchers need the movement to exist so their jobs having meaning when they find out the answers. The solution has to start with the environmental movement because it's only them who can make or break the legitimacy of the CRU, IPCC and UNEP. A few critics here and there won't do any real damage to the billion dollar climate change industry. Real change requires each individual in the environmental movement to honestly look at their actions, their motivations, and the other 11 slices of the pizza. But this is something no one in the movement wants to do because it causes too much pain. To realize that all your efforts in the narrow field of climate change is helping corrupt institutions flourish is a realization that few want in their lives. Most individuals prefer the self-deception that everything is okay with their lives and the world. Few individuals want to look at the world for what it really is and see all the pain and suffering that continues because good individuals do nothing. Why do they do nothing? It's easier. It's more comfortable to take a passive role in life and not have to deal with the hardships in changing things that are bigger than yourself. Paradoxically, this is the initial reason why the environmental green movement exists in the first place—for individuals to be a part of something bigger than themselves—a community that holds the interests of the Earth to a higher degree than their own lives. This is why these individuals don't want to rock the boat even when they know the truth. They don't want to lose the bonds they have. They don't want to be the first person to question the movement and feel the cold shoulder from the others. My advice? To know the truth about climate change and do nothing about it is what government and industry are counting on. Even when individuals know that their actions are not effective (how much of a reduction is there in CO_2 levels globally since 1988?), they don't want to experience the feeling of not affecting events the way they envision it.

This is a difficult chapter for any individual part of the environmental green movement to read and I commend you for doing so. My last point is to emphasize the achievements of the movement, the first being environmental laws. Environmental laws don't exist in the 60's. Oil corporations, the steel industry, and even the war machine government don't care about environmental laws in this time. It's thanks to environmental trailblazers and pioneers who go against the grain—telling billion dollar industries their actions are unethical. It's only because of this constant pressure that various governments around the world make regulations and protection policies such as the Montreal Protocol.

My second and final point is for environmentalists to be proud of their efforts in revolutionizing green building and sustainable housing. Though it's still in its infancy and has lots of room to grow, the field of green building in the 21st century is a major accomplishment. Choosing a housing option that reflects ethics and a natural harmony with the earth in the modern age is no small feat. Growing up in the 90's I don't recall individuals asking whether their house has proper heat insulation, or thinking their roof can generate energy with solar panels, or whether the amount of the water in the toilet bowl is excessive. The new industry of green architecture will have a huge demand over the next few years, decades and centuries to come. And I look forward to watching more YouTube videos on it so when I'm ready to build my own house with my bare hands I'll know what I'm doing.

Chapter 7:

Solar Energy, The Best of all Renewables

The sun is just so generous in all of this energy it rains down on the land day after day after day. It's free and we're not so stupid that we can't figure out ways to exploit that sun.

David Suzuki, The Nature of Things, The Suzuki Diaries.

Now that the preceding evidence proves oil (and other hydrocarbons) to be abiotic, it's necessary to create an analytical model of renewable resources to find out which is the best. Creating this will enable the general public, NGO's and the global environmental green movement to pressure governments for change. *You can't call oil a renewable resource, you only proved that it's abiotic.* Not so. To say hydrocarbons are abiotic is to imply it's constantly replenishing itself by the natural processes in the earth. As long as we have the inner and outer core of the planet functioning properly, and our sun doing what it does every day, hydrocarbons won't run out. But the hypothetical comment is a valid point. Most contemporary authors that are against the peak oil theory or advocate abiotic oil don't go the extra leap and place hydrocarbons in the renewable resources category. Any author with a significant social status will be committing suicide for their careers and be a heretic to do so. But remember, I have no significant social status to lose so the truth will receive a warm welcome as it takes its place on the global stage.

Before I move into the in-depth analytic model, one final annoyance must be put to rest. Can any person (including me) make the claim that oil will last forever? No. The environmental green movement will be quick to point out this flaw; hence, it doesn't deserve to be in the prestigious class of renewable resources. But their excitement only leads to an error in their ability to reason logically. If an individual asks an environmental scientist whether our sun will last forever, the answer will be a definitive no, but always with the claim that it will last a very long time. How much time are we talking about? About five billion years (not including erroneous actions by humanity or unknown cosmic events). The average lifespan is less than 100 years. The average individual doesn't make choices or have incredible foresight in their life within the framework of five billion years. If they do, that's impressive; such a mindset is beyond rational thinking. If you create a model time scale of 1 billion years on a regular sheet of paper, you'll find that 100 years occupies maybe half a centimeter in comparison to a billion years.

But let's go back to our original question, will oil last forever? No. Will our sun last forever? No. But our sun will last a very long time apart from the minute lifespan of humans so it's correct to categorize it as a renewable resource. Oil is in commercial production for only 160 years and coal for 250 years; that's more than the average lifespan but not exactly close to the 6000 years of human history. The sun shining then is the same one shining now. The same is true of oil if commercial production begins 6000 years ago. This is why the general public is unsure about the future of oil and not at fault for their beliefs of oil as a product of vegetation and dinosaur remains. The average individual perceives oil as a finite resource because of generational school and economic indoctrination. Brainwashing without malevolence is still brainwashing.

So, the sun won't last forever but a very long time; so to with oil; it won't last forever but it will last a very long time. Despite the short-coming in attempting to reach the absolute claim of forever, both the sun and oil belong in the renewable resources category.

Analytical Model of Renewables Resources

The renewable resources I include in this model are solar, wind, water, hydrocarbons, geothermal and biomass. I rank geothermal and biomass renewables as newcomers to the modern renewables camp; so they get the 5th and 6th position when I refer to them in general terms. The big three renewables are undoubtedly solar, wind and water; their utilization and resource capability is widespread. In between the newcomers and the big three, hydrocarbons such as oil, coal and natural gas fits perfectly, hence, the title of the book, *Oil, The 4th Renewable Resource*. But this is merely a general ranking without the analytical model showing the true results.

The model I'm using is a point system structure. A list of variables is on the left hand side for judgment via points as to whether a particular resource is beneficial or not. For this analytical model I have 24 variables as the deciding factors. While various categories can exist, I separate these into five: Efficiency, Safety, Social Impact, Environmental Impact, and Susceptibility of Exploitation. Each variable has a value of points, be it 1, 3 or 5 points. The renewable resources gain or lose points as a pass or fail. This means that if the variable has a value of 5 points the renewable resource will either get all five or lose all five. There are no partial credits such as 3/5 or 1/5. The increase in the value of points highlights the significance of the variable. For example, the possibility of death during the building process of a renewable resource is more important than generating energy at minimal start-up costs. This model may sound complex in the explanation but it's quite simple in the modular form. The strength and weakness of this analytical model is dependent on the variables. I'm sure that many individuals can think of more variables. NGO's and professional environmental organizations are most likely capable of creating a model with hundreds of variables. Everyone is welcome to add more variables or use this analytical model as a foundation and retool it. One last thing is that all variables have to measure up against a renewable resource. There can't be any "not applicable" (N/A) symbols. If only one renewable resource can perform a function it shows the strength of this resource and the lack of flexibility and capability of the others.

Analytical Model of 6 Renewable Resources

HC = Hydrocarbons

Geo = Geothermal

Bio = Biomass

Variable	Point Value	Solar	Wind	Water	HC	Geo	Bio
Efficiency							
1. Current Utilization Generating energy at minimal costs.	1 point	+1	-1	-1	-1	-1	-1
2. Current Utilization Generating energy at minimal maintenance costs.	1 point	+1	-1	+1	-1	-1	+1

Variable	Point Value	Solar	Wind	Water	HC	Geo	Bio
3. Current utilization to provide heat & electricity to individual housing units.	3 points	+3	-3	-3	+3	+3	-3
4. Current utilization to provide heat & electricity to city/state/ province.	1 point	-1	+1	+1	+1	+1	+1
5. Current utilization to provide heat & electricity to a whole country.	1 point	-1	+1	+1	+1	+1	-1
6. Current utilization to provide heat & electricity to the global population	5 points	-5	-5	-5	+5	-5	-5
7. Potential to provide heat & electricity to individual housing units.	3 points	+3	-3	-3	+3	+3	+3
8. Potential to provide heat & electricity to city/state/ province	1 point	+1	-1	-1	+1	+1	+1
9. Potential to provide heat & electricity to a whole country.	1 point	+1	-1	-1	+1	-1	-1

Variable	Point Value	Solar	Wind	Water	HC	Geo	Bio
10. Potential to provide heat & electricity to the global population.	5 points	+5	-5	-5	+5	-5	-5
11. Current utilization for multiple applications (residential, work, travel).	3 points	-3	-3	-3	+3	-3	+3
12. Potential for multiple applications (residential, work, travel).	3 points	+3	-3	-3	+3	-3	+3
13. Possibility to provide clean drinking water at minimal costs.	1 point	+1	-1	-1	-1	-1	+1
14. Possibility to enhance food production (greenhouse).	1 point	+1	-1	-1	-1	+1	+1
Safety							
15. Possibility of death during building process.	5 points	+5	+5	+5	-5	-5	+5

Variable	Point Value	Solar	Wind	Water	HC	Geo	Bio
16. Possibility of death during energy generation process.	5 points	+5	+5	+5	-5	+5	+5
17. Possibility of dealing with hazardous material in all phases.	1 point	+1	+1	+1	-1	-1	-1
18. Possibility of accidents & serious injury during all phases.	1 point	+1	-1	+1	-1	-1	+1
Social Impact							
19. Current energy utilization without damage to public health.	1 point	+1	+1	+1	-1	+1	+1
20. Current utilization to build & generate energy individually (do-it-yourself).	3 points	+3	-3	-3	-3	-3	+3

Variable	Point Value	Solar	Wind	Water	HC	Geo	Bio
Environmental Impact							
21. Current utilization without damage to oceans/land/ wildlife during building process.	3 points	+3	+3	-3	-3	-3	-3
22. Current utilization without damage to oceans/land/ wildlife during energy generation.	3 points	+3	+3	-3	-3	+3	+3
Susceptibility of Exploitation							
23. Current utilization without exploitation to individual consumers.	5 points	+5	+5	+5	-5	+5	+3
24. Current utilization without monopolies & corruption of local/state/ provincial/ federal gov't.	5 points	+5	+5	+5	-5	-5	+5

Variable	Point Value	Solar	Wind	Water	HC	Geo	Bio
Total points: **Minus negative values from total points.**	**62 points**	**52**	**30**	**26**	**26**	**24**	**42**

With a significant lead, the winner is solar with 84%; following it at a respectable rate is biomass with 68%; and wind trailing with 48%. I'll explain the results for all six renewables starting from the loser to the winner.

Geothermal

The biggest loser is geothermal with 39%. The general reason for its low score is due to its lack of flexibility in multiple applications. It also has similar problems with that of hydrocarbons such as monopolies and dangerous drilling operations. While geothermal energy can operate locally for housing units, adopting it for a state or national level leaves it prone to environmental damage.

Hydrocarbons

Though it's been in use for 250 years the ranking puts it second to last due to the actions of individuals who squander this resource for their own gain. The major drawbacks are of course monopolies, exploitation to consumers, and environmental impacts. The greatest weakness is from the safety risks and loss of life during the building and energy generation for oil wells, coal mines and offshore oil rigs. The multiple uses in current utilization and potential aren't enough to push it into the top three. Oddly enough the superficial ranking of 4th is the same as the analytical model. Being the same as water, both can rank as 4th or 5th.

Water

Water gets the same rank as hydrocarbons because of its lack of flexibility in multiple applications. A wishful application for water that many hope for is a fuel for cars. It's doubtful such an event will ever happen, but if it does the public will see a strong push for water privatization and increasing water bills. One significant factor in water receiving negative scores is in the Efficiency category. Yes, water can generate electricity but it can't provide heat. If I separate water's utilization to generate electricity and provide heat they'll both cancel each other out, one positive (electricity) and one negative (heat). Some may argue that both heat and electricity shouldn't be in combination, but it has to be together. Solar, biomass and hydrocarbons can generate both electricity and heat, while water can only generate electricity. And yes, the electricity can power electric heaters but it's not a sustainable way of living.

Wind

Wind is similar to water's ranking with a lack of flexibility in multiple applications and has the same electricity/heat problem with negative values. The only reason it does better is due to its light environmental impact. It may not be aesthetically pleasing to look at but it doesn't displace wildlife as huge reservoirs do for hydroelectric plants. Some will argue that birds and geese die from the blades, but it's no more significant than birds hitting massive skyscrapers in the cities.

Biomass

The high ranking for biomass (in comparison to the previous four) is due to its multiple usage in providing heat, electricity and fuel for transportation. It's even capable of filtering dirty water in a bio-sand filtration process without any electricity.

Biofuels are incredibly expensive and not cost-effective, but can be a direct replacement for hydrocarbons. Producing biomass commercially does require large infrastructure but can also be done as a do-it-yourself project for heating and vehicle fuel. The industry's ability to exploit consumers is low due to the fact that consumers can make their own biofuel if it's too expensive at the pumps. The lower ranking in comparison to solar is due to the environmental impact it can have. Creating biofuels at home doesn't displace any wildlife or land management. However, the biomass industry is growing and large tracts of land for growing bio-crops are taking up significant space. This is not about a few local biomass companies providing biofuels, instead, it's about large global biomass corporations that specialize in producing and exporting bio-products. While biomass it's good at providing individual or local needs, it's currently unable to provide extensive support for a whole country. It can provide electricity for the grid but unable to do so for individual housing units. Yet this may change in the future. With the exception of solar energy, biomass' potential out of the five to provide new forms of energy generation is quite extensive.

Solar

The big winner is solar due to its flexibility in usage. The ability to use it in do-it-yourself projects at the individual level, and the inability to monopolize the sun is quite significant. The incredible energy generation potential for solar energy on a mass scale is what gives it the highest marks. The sun is not going anywhere for another 4-5 billion years so the global general public has quite a long time to utilize it as a primary source of energy. The reason why it loses a few marks is because of the stagnant solar photovoltaic (PV) cell industry. The costs are coming down for PV's but the concept is incorrect. Photovoltaic cells transform light from the sun into direct electrical current (DC). This is a good beginning for the solar industry but that's where the compliments end. PV cells are one of the most inefficient concepts. The fact that only bright solar radiation (a clear day) will enable a charge is its greatest weakness. PV cells are unable to utilize the tremendous potential of the sun's energy. Million dollar companies that specialize in photovoltaic cells are actually holding back the solar industry. All PV cells produce voltage but it's unreliable with various high draining electrical devices. Concentrated Solar Power (CSP) is different from photovoltaic cells because it uses mirrors and lenses at different curvatures to focus sunlight in one area. The sunlight converts to heat in order to power turbines, thereby creating electricity. Spain is the world leader in CSP with its solar tower plants. CSP technology is still at a beginner level in terms of utilizing solar energy but it leans more towards the intermediate level with its ability to store heat energy when there's no sun. This in turn can create a constant stream of electricity for various uses. While PV cells work for individual houses, CSP can power cities/states/provinces. The limitations of both concepts are their dependence on clear bright sunlight. The solution is in new innovations.

In September 2008 a news report mentions a teenager creating a 3-dimensional solar cell that's capable of absorbing visible and ultraviolet radiation (UV).[1] Currently, all commercial PV cells and CSP's are 2-dimensional cells that can only absorb visible light. A 3D solar cell revolution from this teenager is capable of significantly utilizing solar energy on a much more efficient level:

> *In his project, 'A Highly-Efficient 3-Dimensional Nanotube Solar Cell for Visible and UV Light,' William invented a novel solar panel that enables light absorption from visible to ultraviolet light. He designed carbon nanotubes to overcome the barriers of electron movement, doubling the light-electricity*

conversion efficiency. William also developed a model for solar towers and a computer program to simulate and optimize the tower parameters. His optimized design provides 500 times more light absorption than commercially-available solar cells and nine times more than the cutting-edge, three dimensional solar cell.[2]

This work alone is promising but solar innovation is widespread. At MIT Dr. Andreas Mershin, a physicist, is innovating new methodologies from Earth's natural solar panels...leaves:

Here is the recipe his team discovered: organic matter, like grass clippings or agricultural waste, is mixed with a stabilizing peptide powder and from that protein is extracted from the materials, stabilized. Protein is essential for photosynthesis. This mixture is then spread on a glass or metal surface and through photosynthesis, it keeps on generating electricity.[3]

This process is very different than 3D solar cells but both are pushing the intermediate level in utilizing solar energy, and finding new means to accumulate, store, and use energy in multiple applications.

These innovations are excellent in solar energy development but still pale in comparison to planetary applications. One barrier that exists to utilizing solar energy is the various filters it passes through in our atmosphere, and of course the usual cloud coverage. One possibility is to build a tower with a 3D solar array above the clouds and absorb both visible and UV rays. Clouds form at 10-30,000 feet so a structure higher than that is necessary. It may sound like a pipe dream, but humans are quite capable of building megalithic structures. And if only a few are necessary across the planet for perpetual energy, it can be done as an international project. One final solar innovation that can force visionaries and engineers to stop thinking stationary is to build structures in space (outer atmosphere), and make use of the sun directly without any filtration. This is going beyond the simple visible and UV rays and moving into the concept of cosmic rays. As fanciful as it is, such projects don't warrant an immediate rejection due to a lack of technological capability, or a narrow perception of growth. A perceptual framework that takes into account the global population beyond national boundaries is what will lead to real success in utilizing massive amounts of solar energy.

Chapter 8:

What You Can Do About It

For the great enemy of truth is very often not the lie--deliberate, contrived and dishonest--but the myth--persistent, persuasive, and unrealistic. Too often we hold fast to the clichés of our forebears. We subject all facts to a prefabricated set of interpretations. We enjoy the comfort of opinion without the discomfort of thought.

US President John F. Kennedy, Yale University Commencement Address. June 11th, 1962.

Just as most philosophers lack the ability to play an active role in citizenship, the general public as a whole prefers to act passively for many contentious issues. That's not to say there aren't hardcore demonstrators left, there are. But the larger middle class of activists that tread lightly on government corruption overshadows them. Non-profits are trying to bolster this lack of public momentum by rallying them behind various causes. Yes, non-profits and NGO's working in 3rd world developing countries to alleviate poverty is good. Yes, creating organizations to fight social injustice locally and internationally is good. But what are the results? Africa still lingers as a 3rd

world developing nation despite 50 years of admirable efforts. Social injustices such as corporate greed, corruption and discrimination are as prevalent today as it is a hundred years ago. Why? Two reasons. The first is due to the framework of the Western world. The reason why there's so little progress from individuals with good intentions is because the economic and political framework's core is "profit at all costs." Policies on the surface usually advocate for one thing, but underneath, the implementation never lines up. How many policies exist in the Western world where implementation is ineffective or non-exist? It's not that policies can't work, but rather, it may not be "in-line" with the current economic and political framework as a whole. The results equate to inefficiency, ignorance and unaccountable representatives.

The second reason why most actions lack the necessary power to cause change is due to the global general public. In regards to 3rd world developing nations, the hard truth to accept is that the global general public as a whole doesn't care enough to take action. They care more about their triple cappuccino they get at the Tim Horton's drive thru (the Canadian equivalent of America's Dunkin Donuts). They worry about getting to work on time before their boss gets in a bad mood. They care more about suing an individual that causes them emotional duress then the lives of starving Africans. If a quarter of the global general public focus on ending 3rd world poverty they can achieve it in a matter of weeks; the reason it doesn't happen is because private interests rig the game. The global general public, consumers, students and parents make decisions within the Western framework of duplicitous agendas. And those that try to "think outside the box" usually get their financial world topsy-turvy. And it's personal finances that demonstrate the realism of any situation no matter how strong the conviction is. Do you really think corporate executives, United Nation officials or federal politicians ever worry about paying their rent? Having enough money for food and gasoline? Or whether their child will get a proper education? Of course not. Such matters only exist in the global general public arena. And these simple issues cause misdirection in the focus of their lives. Generally speaking, the public knows all of this. These words are nothing new to the equation. The global general public just don't like to hear it because it's too much truth to handle.

And where does oil fit into all of this? Remember, oil is the lifeblood of the Western economic and political framework. Take action that can affect oil policies and everyone will perk up to listen. The best action to take upon reading this book is to focus on the education system. Millions of students and parents in North America and around the world have an advantage over me. They can see and go through the actual science textbooks in the schools. I only present six textbooks in Chapter 2 out of millions of others that circulate in the schools. Parents are in the best position to demand local, state and federal education departments to change their curriculum in regards to the false information about fossil fuels. Parents and students especially, can also go to their schools and give the teacher a copy of this book. The evidence in Chapter 3 is enough to prove the myth of fossil fuels to be incorrect. If that's not enough, send one to the principal and the superintendent. Ask them to rebut the claims that I make. Are their answers believable? If enough students and parents inquire about the myth of fossil fuels to teachers and principals, a debate will follow and the initial seed of doubt in the fossil fuel theory will begin. University students can do the same by going to their professor after reading this and ask about the creation of oil. If the answer isn't satisfactory, show them another theory in this book. Acting at the local level (high schools and universities), is the first step. The second step is the state education department. All local jurisdictions implement policies that come from the state/provincial level. This means that each grade and subject matter has a curriculum guide. The science and geology curriculum guide will show what students are learning. And the myth of fossil fuels are definitely in there.

Here's a sample of the California science content for public schools:

> *Students know the global carbon cycle: the different physical and chemical forms of carbon in the atmosphere, oceans, biomass, fossil fuels, and the movement of carbon among these reservoirs.*[1]

And you can bet twenty bucks the definition of fossil fuels in every school textbook is a compound that comes from vegetation and animal remains. Talk to your state department education officials, write letters and send emails. Don't let ignorance prevent you from calling into question the myth of fossil fuels. One reason many students or parents will shy away from such actions is because of embarrassment. No one wants others to look at them as unintelligent. If you can get over this barrier you can handle the rest.

And the final step in causing change in the education system is to speak to your local political representative. This is the highest you can go in advocating change in the education system. You'll have to be firm with letters or speak to them in person. Politicians in general don't act on a matter unless there's constant pressure by follow-up calls, letters and emails. The best way politicians avoid taking action on contentious issues is to bet on the general public losing interest. As long as there's constant pressure from numerous citizens, politicians can't sweep it away, and curriculums have to change to correct this long stubborn error.

Education textbooks can always change but the implications of this book are far more reaching than just that alone. Enabling the public to rethink the national policies on oil, buying it from other countries, and how to utilize it properly, is another reason why this book exists.

Nationalizing Oil?

Nationalization is a contentious issue in the US, and any changes in policy will likely have a domino effect in other countries. Nationalization isn't new though; it's done in many Middle Eastern and South American countries. Nationalizing oil means the federal government will own, mine and operate oil production and refineries with full control. There are no private corporations that decide on the quantity of oil production; nor do they share in any of the profits. Argentina is the newest country to nationalize oil from private hands. But is it more effective? What are the economic results in these countries? Whether the oil production process is more efficient in these countries is debatable. And a template nationalization model doesn't necessarily apply to other countries due to unique economic realities. But the economic results can be monumental. The Middle East is a desert with small villages and numerous tribes before striking oil. Though such areas do still exist today it's by choice, not lack of dollars. Anyone who goes to the Middle East will see the luxurious lifestyle of government officials and basic Western amenities such as shopping malls, eateries and amusement parks...all on par with their Western counterparts or exceeding them. The largest oil producer according to EIA's 2011 production statistics is Saudi Arabia with 11,153 barrels per day; and then Russia with 10,228; and finally the US with 10,107.[2] But the nationalization of oil in Saudi Arabia, Russia and Brazil are all national priorities and a part of their economic policy. This isn't the case for the US as one of the largest players but sitting in private interests.

So the question remains, is nationalizing oil an economic priority in North America? Maybe. More importantly, can oil nationalization even happen in the first place? Definitely not. The roots of oil go back 160 years and that's where we find the answer. The first oil well starts up in 1859 and put into full commercial production throughout the 60's. Rockefeller takes a monopolistic approach to oil in the 1870's and this continues in the present. A hypothetical scenario will provide a clear answer.

Let's say J.D. Rockefeller isn't in the picture—he doesn't exist at all in our hypothetical scenario. What happens to the oil industry? Does a monopoly on oil exist? Absolutely not. *How do you know?* Because from 1859-1869 there are no oil monopolies or merging refineries, they're all independent. Some are much larger than others but all compete fairly trying to get the best rates in buying oil and shipping it. It's not until Rockefeller comes into the picture in the 1870's with Standard Oil that oil monopolies begin to form. But after he becomes a significant player, opposition is immediate from the independent refiners. The only reason why the Producer's and Refiner's Union exist is due to Rockefeller. All throughout the 41 years of Standard Oil's creation, independent businesses keep fighting his oil monopoly. Unfortunately they're not successful and Standard Oil absorbs them or runs them out; but their desire for independent existence and fair competition never wanes, despite it being a losing battle. And this is what bothers me, without J.D. Rockefeller a monopoly on oil doesn't exist and no one else has this manipulative desire. So how is this man in the right place at the right time of the oil boom? It's almost as if he's put there by some unknown means. Are psychics telling his parents, grandparents or great grandparents to go start a family in Cleveland with the promise of enormous wealth? Any answer is speculation; but if Rockefeller is not in the picture the possibility of nationalizing oil in the 1900's exists. The possibility can also arise during WWI under a presidential order, or the crash of the 1920's, or during WWII. The fact that it doesn't happen during any of these moments is significant. 65 years later North America's economic policies still make nationalizing oil impossible. But more importantly, the oil corporations that run the private sector government will never allow oil to be under the control of the bureaucratic public sector government. Trillions of dollars in the past, present and future is not something oil corporations will hand over to someone else.

Nationalizing oil in North America is not possible within the current economic framework. But there's no harm in asking whether governments want to nationalize it. The answer, like the first, is a definite no. *How do you know for sure?* Looking back at our hypothetical scenario, the government's decision to leave oil in the hands of the private sector, when it has three chances to take it, demonstrates an intentional action. The US or Canadian government doesn't want to nationalize oil. Can it be they want to but unable to? Nope. Definitely not. Remember from Chapter 5, the US government never stops Standard Oil's unethical practices until the independent businesses are practically begging. They condemn it publicly in the 1870's but take no action. And when they do step in all indictments and court hearings don't advance because the government withdraws the charges. The 1911 break-up of Standard Oil into 34 companies is a joke. A few years and decades after, the companies absorb others and merge themselves together. The modern results are BP, ChevronTexaco, ConocoPhillips and ExxonMobil. These four corporations are separate entities under the law, but it's also correct to call all four of them Standard Oil. And this is what exists today—Royal Dutch Shell and Standard Oil. Yes, there are many other oil corporations such as Russia's Lukoil, and the Brazil's Petrobras. While these corporations are big players, the fundamental difference is the nationalization of their oil; a very different league than the private oil sector.

Will the US and Canada nationalize their oil in the foreseeable future? Nope. The reason is very clear. They don't want to. *But why not?* Because it's too profitable keeping it in private hands. The government knows it's better to let the private sector have control of the oil industry, and then take a large cut of the profits while doing absolutely nothing. The Canadian government doesn't have a federal royalty rate for oil because most of it is in Alberta. This equates to a provincial royalty rate of 40% for oil and 36% for natural gas.[3] The US government has a federal royalty rate of 18.75% for offshore and 12.5% for onshore (which they're attempting to increase).[4] The US government also has state royalty rates done separately by each state. It's not accurate to compare the two countries rates because one is for a province and the other for a country. But for both governments the amounts are in the billions, supposedly going to various infrastructure projects (whether it really does go towards this is questionable). But the royalty rate alone (allowing oil corporations to drill on their land), is nothing in comparison to the fuel taxes it collects. The first fuel tax in Canada begins in Ontario at three cents a gallon in 1925.[5] The US government beats them 6 years prior with Oregon charging one cent per gallon in 1919.[6] All provinces/territories/states and federal jurisdictions follow suit. On the surface a one cent tax in the US doesn't seem like a lot but it accumulates over time and increases by 10-73 cents a gallon, equaling trillions of dollars. This is the most efficient strategy the Canadian and US government is using to fund infrastructure and capital projects.

Realizing how much money comes from doing nothing, these governments will never nationalize oil because of the profit. But where are these profits coming from? Government royalty rates come out of an oil corporation's pot either in cash or oil. And it's the general public that pays for all local and provincial/state taxes. The oil corporations lose a bit but recover the royalty funds due to tax breaks. And for the government? They never lose. The consumer pays them through fuel taxes on every liter/gallon and they get their royalty cut from the oil corporations. There's no way the government is going to change this type of arrangement by doing something stupid like nationalizing oil.

Chapter 9:

Answering Critics

1. *Oil fractionates as 275° C so if a hydrocarbon like oil is coming up from deep within the earth it will be completely disassociated as it reaches the surface.*

Oil fractionation means that hydrocarbons become unstable as they rise to the surface and gradually lose their hydrogen atoms due to oxidation. If they become unstable, hydrocarbons such as oil can't form. This objection comes from oil being put into high temperatures in a lab and seeing it fractionate. The error in this criticism is a lack of understanding of pressure and temperature. Yes, the oil does separate in a lab, but the results are due to faulty methodologies. The pressure in the lab is 14.7 psi (standard surface pressure of the earth). A temperature of a few hundred degrees at a pressure of 14.7 psi does lead to fractionation. But placing the oil in an airtight container with a pressure of 35,000 psi for example (similar to 8 km in the earth), and left for hours at a few hundred degrees, the molecules will stabilize and re-organize themselves, still maintaining their hydrocarbon characteristics. Gold explains it more clearly:

> *To get a sense of how hydrocarbons spontaneously reconfigure their molecular mix under a substantial change of pressure and temperature, I directed a graduate student to study chemical changes in a sample of propane (C_3H_8) subjected to a simulated depth environment of 475°C temperature and 4000 atmospheres pressure, which corresponds to a depth of about 10 kilometers in much of the earth's crust. After only six hours, the sample had rearranged itself into a mix ranging from C_1 to C_5, maintaining the input ratio of carbon to hydrogen.*[1]

Pressure and temperature are the key components in answering this criticism. But other factors such as the speed of the hydrocarbons moving upward, the length of time it passes a given space in the earth, and the chemical properties itself...all play a factor. The real problem that chemists, geologists and geophysicists face in proving or disproving the abiotic theory is not so much the data and observations, but rather, the ideologies of how the earth operates at deep subsurface levels:

> *At a depth of 150 kilometers, for example, the pressure would be 40 kilobar, which is equivalent to 40,000 times our atmospheric pressure. Very many different molecules will be held together by that level of external pressure—molecules that we have never seen at the surface. In fact, the very concept of molecules begins to break down at pressures comparable to the forces within*

a molecule that hold it together or make it come apart.[2]

At these deep levels in the earth the 4th state of matter exists: supercritical fluids; a mix between a gas and a liquid. It can be done in a lab and is a good project for science students in high school. An example that Gold uses is water. At sea level water boils at 100°C. But almost a kilometer deep in the ocean the boiling point changes to over 300°C because of the pressure.[3] Students need to remember this lesson well. The next time your teacher says water boils at 100°C, correct them and say it boils at that temperature only at sea level. If the depth changes so does the pressure; which means that the 100°C general answer is incorrect.

2. *Coal has biological peat and lignite material in it, this holds true with the definition of fossil fuels.*

Like the first criticism, the second looks at the end result and not understanding how it gets there. If a student writes the answer for a difficult mathematical equation without showing how they get the answer they'll lose marks. The process the student uses to solve the equation is necessary to understand the problem. Scientists look at coal on the surface and say it has peat and lignite (biological material) decomposing. This implies that coal comes from swamp vegetation. The key to understanding this is to remember coal (a hydrocarbon) can form anywhere on the earth, and right under peat and lignite fields are no exception:

In this peat field, the gases just below the surface were greatly enriched with methane, as is commonly the case in such environments—a condition that supporters of the biogenic theory of course attribute to the presence of methane-excreting bacteria feeding on the

> *plant debris in an oxygen-poor environment. But the gases were also enriched with all the other hydrocarbons gases from C_2H_6 to C_5H_{12}. This mix is not normally produced by plants in any stage of decay. Quite simply, microbes do not excrete pentane while decomposing carbohydrates.*[4]

This example shows how methane (natural gas) is usually seen in high concentrations in coal mines. In fact, many fires and hazardous accidents happen in coal mines because it's so rich in natural gas seeping up from below. Here are two common layouts of the deposit formation for hydrocarbons.

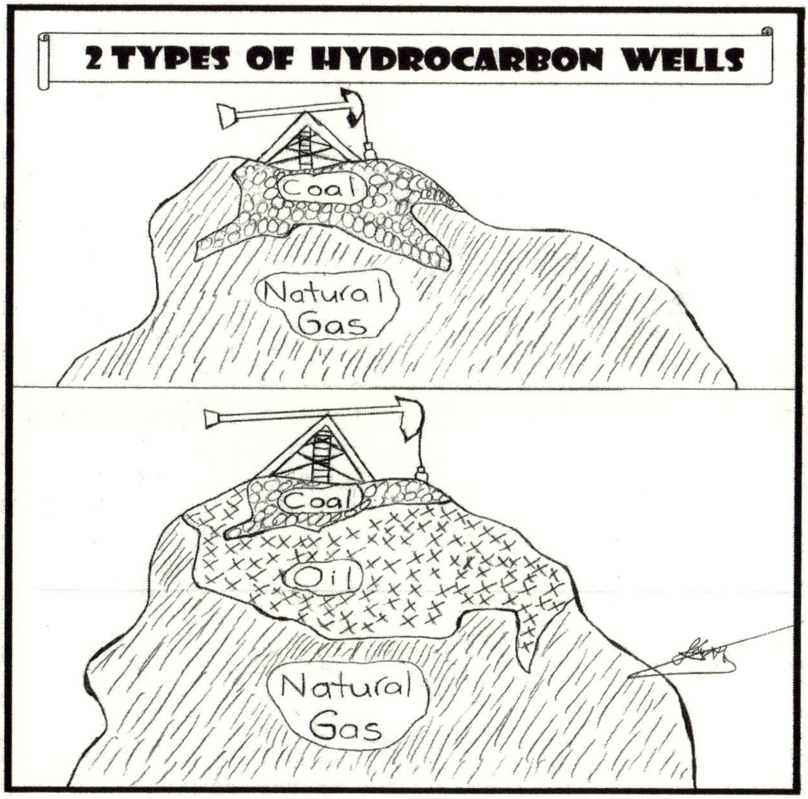

The mining process subdivides coal into various types but two are sufficient to understand the difference: black coal and brown coal. Black coal is a result of purely abiotic processes where the hydrogen is susceptible to oxidation and leaves only the carbon coal. Brown coal is the same process, but as it comes to the surface it mixes with the peat and lignite in a given area. This means that brown coal is the result of both the abiotic and biogenic process.[5] It should be clear that while brown coal is the result of both processes it can never be the result of a purely biogenic process. This means that the biogenic theory falls apart for both criticisms.

How Oil Corporations Will Respond

There are very few times oil corporations have to respond to an issue. Be it fraudulent accounting numbers or a disastrous spill, the first rule is no comment; second, to deny it; and finally, blame someone else. Sure there are fines, but many companies pay it as if it's nothing. In 1998 I remember General Electric (GE) shelling out a few hundred thousand for environmental damages.[6] Does it hurt their profit? You tell me. In 1998 their profit is $9.2 billion, in 1999 $10.7 billion, and in 2000 $12.7 billion.[7] In 2011 their profit is $14.1 billion.[8] I'm not seeing any real financial damage. Can you see any? The problem is that any environmental fine a corporation pays does nothing to hurt their bottom line; nor does it create an incentive to stop their polluting actions. In fact, it's easier and cheaper for a corporation to pay an environmental fine than it is to pay for proper disposal or enforcing environmental regulations. Fines are not the answer. Real accountability is. When corporations are flagrant in their abuse of the environment, government policy in revoking their charter/license is necessary. This action is capable of causing real change for multinational corporations to be environmentally responsible. Revoking BP's charter is what the US government needs to do in the aftermath of the oil spill. Such actions will send a clear message to oil corporations and corporations in general. But alas, governments are either unwilling or unable to take such dramatic actions. Only the global general public can push their governments to become better than their lazy bureaucratic process allows them to be. Only then are they worthy of the title of a democratic government.

I'll create a fictitious hypothetical scenario showing a possible response due to this book and the public's reaction.

Nathan: Oil chairman representing all private oil corporations

Danielle: Marketing representative

Jerry: 2nd in command

Paul: Oil executive

USA, February 10th, 2013. Nathan arrives at home after a long day. The phone rings...

Nathan: Hello?

Jerry: Nathan you need to come down to the office right now.

Nathan: I just got back Jerry. Is one of our rigs leaking?

Jerry: No it's nothing like that, it's completely different, something unexpected.

Nathan: *Something unexpected? Jerry is the best analyst in the company. No one can anticipate government and economic forecasts better than him.* Fine...I'll be there in 20.

At the office, Jerry throws down a copy of 'Oil, The 4th Renewable Resource.' Nathan skims through it in 5 minutes.

Nathan: This is what you called me down here for? This is a nothing. It tells the truth about oil but who's going to believe this. It probably sells on the discount book shelf for a dollar.

Jerry: It's selling over a million copies. People are starting to ask questions.

Nathan: There's asking us?

Jerry: No, they're asking the government to change all science textbooks in the US. The government is asking us what action is advisable.

Nathan: Order some food and give me a few hours to read this.

Jerry orders the food and leaves him alone with the book.

Nathan: *What the hell...so many connections: abiotic theory, hydrocarbons in space, Standard Oil, school textbooks, media scares, our connection to the climate change industry, the Rockefeller foundation reports, the UN and government. Who is this guy? Why now? We've been too careless in our actions, overplaying our hand, overconfident that nothing can stop us.*

Nathan asks Jerry to come back.

Jerry: Well?

Nathan: What does the internet say about him?

Jerry: He's a nobody. A Canadian author, works with Aboriginal youth, does his schooling in Toronto majoring in philosophy.

Nathan: A Canadian author? Yup...no chance of seeing this one coming. Don't feel so bad Jerr, no one can anticipate this.

Jerry: Sorry. What do you want to do?

Nathan: No need to go heavy yet. Give him a shout and ask him to explain his book to our staff and executives. We'll get pictures when shaking hands and we'll give him some donations and feed it to the media. The guard dogs will tear apart his character and make him seem less credible. Standard protocol.

Jerry: You got it.

Two weeks go by. Jerry enters Nathan's office.

Nathan: An update on our unique problem?

Jerry: He refuses to meet. We've even offered him a 5 star resort, all expenses paid trip to the Caribbean to do the presentation.

Nathan: Smart kid, he knows what we're trying to do.

Jerry: Congress wants to meet with us, they're getting too much pressure from the public. Even some politicians are calling for the changes in the textbooks. And the mainstream media are beginning to back the issue. What do you want to do about the kid?

Nathan: He's not the problem anymore. The kid is trying to bring out something that can never see the light of day...Call a meeting for tomorrow. Get the heads of the oil corporations and tell Danielle to come as well. Do it quietly.

Jerry: What are you planning Nathan?

Nathan: Something unthinkable.

The next day five executive directors, Jerry and Danielle attend Nathan's meeting.

Nathan: Thank you all for coming and please forgive the secrecy. It should be obvious to everyone this meeting doesn't exist in the records. *All cell phones are sealed in a magnetic container, the room's been swept for bugs, no paper for them to write on. No records. There can't be any records for this plan.* I'm sure you've all heard about the abiotic issue from the media which is making its way around congress. I know some of you may think the issue isn't big, but if congress gives in and changes school textbooks, it will be the beginning of the end for us.

The representatives fidget in their smooth thousand dollar leather chairs.

Paul: What are you proposing?

Nathan: We give the public and the world what they thought will never happen, an energy crisis and a massive oil shortage.

Paul: How?

Nathan: Through our friends in the public sector. We'll use HAARP. The High Frequency Active Auroral Research Program in Alaska.

Most of the participants share confusing looks with each other.

Paul: How is that going to do anything, isn't it some big antenna that puts high frequency radio waves in the upper atmosphere?

Nathan: It is, and more. In 1992 HAARP is developed by the US Air Force, US Naval Research, DARPA and BAE Systems. It's almost boggling of how the public buys the surface story with such obvious military backing.

Jerry: Isn't there a UN ban on weather modification agreed to in 1977?

Nathan: There is indeed, and there are over 150 weather modification programs in the world today. But I don't see anyone telling China to stop using the chemicals. It's almost scary how mainstream media directs the public's attention in one direction but not in another. HAARP is able to cause earthquakes, droughts, hurricanes, floods and any other biblical plagues.

Paul: So how are we using this to cause a shortage and energy crisis?

Nathan: The energy crisis comes first. We'll get our friends to target electricity generation plants throughout the country that rely on hydro, coal, a few of our diesel refineries, and the usual coastal areas in the south; and of course no one can blame nature for any of this. Danielle, you'll use your media contacts to focus on the climate change industry, but tread lightly, we're not using the full force of the media yet. The climate experts will lay the blame squarely on human activity. This won't be enough to change mass opinion but it will cause doubt in refuting the global warming theory. The US will face an energy crisis like none other, forcing them to rely on nuclear and diesel generation. And since diesel has been hit as well, the costs have to increase with skyrocketing demand.

The next stage will happen about one month after the energy crisis.

Paul: Why wait that long?

Nathan: It'll take time to re-establish electricity generation, and nothing is more annoying to the public then getting everything back on track only for something worse to come along and ruin it. This is where we create an oil shortage. The target will be a handful of our refineries.

Paul: Our refineries? We'll lose money!!!

Nathan: Money that's insured against acts of god may I remind you. The refineries have to be hit because the public isn't gullible enough to think our supply of oil is running out anytime soon. And when the plan is ready we can't slow down refining, otherwise people will start to suspect something. It has to be a direct hit from earthquakes and hurricanes using HAARP to throw our refineries into turmoil. Danielle, this is where you'll use the full strength of the media to show the oil refinery devastation and drive home the fact that oil is now scarce. There won't be enough refined oil to sustain the American lifestyle temporarily. The myth of scarcity will no longer be defined as a myth but a result of human activities.

The price of refined oil will skyrocket, maybe $200-500 a barrel. Prices at the pumps will increase three fold and consumers will be filled with anger, but not at us. We've been hit the hardest and have to raise the price, otherwise we can't do business. This is one of the key things to remember. The public is very understanding. As long as they understand our losses and having to raise the price, they'll cope with the situation.

Jerry: Won't the public complain so much and ask the government to step in?

Nathan: They'll ask, but the government won't. America is built on capitalism, free enterprise and fair competition. Changing that is equal to changing the constitution, it'll never happen.

And aside from our huge profits for the next few years in rebuilding our refineries, the issue of abundance in oil will lose flavor with the public, and we'll constantly remind them of scarcity so they'll never try a stunt like this again.

Final Thoughts

Foreign policy is a dangerous term in the US and conjures many different scenarios. In Canada the words have no effect. If you walk up to a Canadian and say "foreign policy," the response will be, *oh…we don't really have a foreign policy. Perhaps you might like an iced cappuccino from our Tim Horton's and a bottle of maple syrup for the kids?* Canadians are more or less indifferent to national foreign policy because it doesn't have ripple effects as the US does. The US foreign policy on oil depends on the myth of scarcity. And this is where it gets interesting. Looking back at Vietnam there are many reasons why the US government participates. But one reason, not necessarily a big one, is for oil. In February 1975 Mobil (later known as ExxonMobil) strikes oil on the coast of South Vietnam. If everything goes according to plan oil rights will stay private and oil nationalization won't happen. It doesn't happen this way and South Vietnam falls, forcing the US government to leave. But it's interesting to look at possible motivations of why the US government stays there for so long. Yes, the surface reason is to battle communism. But a major reason for Western governments to battle communism is to fight oil nationalization.

Going back to 1953 in Iran the US foreign policy enables a puppet government that's favorable to oil corporations, giving US oil monopolies 26 years of profit. In 2001 the occupation of Afghanistan begins under the label of fighting terrorism. But a report in 2010 confirms the existence of massive amounts of copper, gold, lithium, iron and cobalt in Afghanistan.[1] This is enough to sustain Afghanistan and US private interests for generations. And finally, in 2003 the invasion of Iraq begins for various reasons but oil and mineral rights go to US and other multinational corporations. And despite the pullout of troops in 2012, oil corporations are staying. This is not just one or two corporations, but a whole slew of them: Royal Dutch Shell, ExxonMobil, BP, Russia's Lukoil, Brazil's Petrobras, China's National Petroleum and France's Total SA. The Al Jazeera article points out this discrepancy:

> *'Prior to the 2003 invasion and occupation of Iraq, US and western oil companies were all but completely shut out of Iraq's oil market,' oil industry and analyst Antonia Juhasz told Al Jazeera. 'But thanks to the invasion and occupation, the companies are now back inside Iraq and producing oil there for the first time since being forced out of the country in 1973.'*[2]

The Chinese government is also a big player in Iraq's oil industry with almost half of Iraqi oil going to them.[3] The simple reason is because they accept lower margins of profit relative to their Western counterparts.[4] On the surface this may seem counterproductive to the US government invasion, but it actually works in their favor. Essentially, the Iraqi government uses the oil funds (which the Chinese government converts to USD) to pay for US contractors working in Iraq for infrastructure and defense. From the perspective of a conspiracy theorist you can say that the Chinese government is lining the pockets of the US government and US corporations through its Iraqi oil payments.

None of this is mind-blowing of course. The global general public is undoubtedly aware of the private monopolies and exploitation of oil resources across the world. But it's a matter of ethics. Even negating the question of what right one country has to interfere with others, the issue is still about the ethics and fairness to people, be it the Iraqi citizens or North American general public. I may not know all the pieces of the puzzle, but I know that sharing these profits with the Iraqi citizens is not a reality. You can blame oil corporations for the profits, the US government for not doing anything about it, and yourself for choosing to turn a blind eye towards this. What can one person do against a billion dollar corrupt oil industry and government? Everything and nothing.

One problem after reading this book that becomes apparent is who to trust for valid information. Schools are teaching what they're told to; mainstream media organizations and Western governments have a political and economic gain by advocating the myth of scarcity; and oil corporations have a vested financial interest in how all of this plays out. So who do you trust? I can offer one resource from my university philosophy lectures. *Oh dear god no.* I promise this will be short, sweet and simple to understand. It's not a matter of who you should trust but rather how an individual perceives the information. Standpoint Theory (ST) is a philosophical methodology to sort out objective information. The creation of ST is by the German philosopher Friedrich Hegel in the 1800's to understand the relations between a slave and their master.[5] It's not until the 1980's it becomes a methodology for the feminist movement to argue against the male domination of political and scientific doctrine. It has multiple uses today to help a person understand what information they can trust and what goes in the garbage. Standpoint Theory argues that all social relations or affiliations we have shapes are knowledge,[6] such as our political, social, religious, cultural, financial and economic relations. For example, M. King Hubbert advocates the peak oil theory through the Hubbert Curve. Should you trust this information? Using ST we just need to look at his relations. Working for Royal Dutch Shell he has a vested social and financial interest in advocating his theory. His career or social status benefits by advocating peak oil theory. His financial situation and the corporation's financial interest benefits from his claims. Taking these two standpoint relations into account his information regarding peak oil theory is not trustworthy, hence, you discard it. But this doesn't mean you discard all of his knowledge. Working in the oil industry, Hubbert knows the dynamics of oil corporations and various oil recovery methodologies. This information is valuable and trustworthy. It's not a matter of discarding all the knowledge an individual has when one piece is in doubt; but whether that piece of information benefits the individual in regards to their social, economic, or political status. The next time someone tells you something passionately about a cause or issue, use ST to look at their relations and whether they benefit by advocating such claims. And feel free to extend the relations beyond the six I mention. Gender relations, working class relations and managerial structures also apply. And it's not necessary to take a broad approach; narrowing down the focus can greatly help. For example, looking at information from US President Barack Obama, you can see his political relations more so then simply representing the Democrat party. You can look at the right or left leaning group of the Democrat party; or the pro-health care left leaning group in the Democrat party. There's no end to how deep and useful this approach is to gaining objective information without the usual bias.

This is a lot of information to digest so please take your time and mull it over. Give it a few days and see if it makes sense to you. Go back and read it again if you need to. I purposely re-read books about a year after the first read. In that amount of time my perception and character change ever so much, with new passages standing out and new connections forming. The human mind is truly a wondrous thing, and how an individual learns, how they grow, how they understand things...is truly remarkable.

If you want to know more information about the abiotic theory and all the connections in this book I implore you to actively search for it. If you're passionate in the science behind the abiotic theory I recommend Thomas Gold's, *The Deep Hot Biosphere: The Myth of Fossil Fuels*. If you can read Russian texts look into Dr. Kudryavtsev's books. If you want to learn more about the climate change industry I recommend Larry Bell's *Climate of Corruption: Politics and Power Behind The Global Warming Hoax*. If you're more into the international political aspects of the abiotic theory I recommend Jerome Corsi's, *Black Gold Stranglehold*.

If, after reading *Oil, The 4th Renewable Resource*, you still have doubts about the scarcity myth and fossil fuels, please remember that energy is beyond creation and destruction, it merely transforms into one manner or another. The $60.00 worth of gas you put into your car doesn't go to waste in the least. The gasoline transforms via artificial and natural processes. CO_2 leaves the tailpipe and the planet is able to transform and utilize it. You don't need to tell it what to do with the CO_2 or dictate how the planet should operate. The Earth knows what it's doing. You're only on this planet for a short amount of time so make the most of it and have a healthy respect for it and the people around you. But always be mindful of your social status in comparison to the planet. You're a guest, not the owner of planet Earth. It's best not to confuse the two.

Notes

Introduction:

1. Gold, Thomas. *The Deep Hot Biosphere: The Myth of Fossil Fuels.* New York. Copernicus Books, 2001. p. ix.

Chapter 1:

1. Tarbell, Ida M. *The History of the Standard Oil Company.* New York. McClure, Phillips & Company, 1904. p. 5.

2. *Merriam Webster's Collegiate Dictionary* 10th edition. Massachusetts. Merriam-Webster Inc. 2002. p. 459.

3. Petroleum Formation and Occurrence. Lecture Notes at *Massachusetts Institute of Technology.* 12.158 Lecture 11. http://ocw.mit.edu/courses/earth-atmospheric-and-planetary-sciences/12-158-molecular-biogeochemistry-fall-2011/lecture-notes/MIT12_158F11_lec11.pdf

4. Bartlett, Bruce. Oil unlimited? *The Washington Times.* June 8, 2004.

5. Neumann, Casper. *The Chemical Works of Caspar Neumann*, M. D. Trans. & ed. William Lewis. London, 1759. 2nd edition 1773. Vol. 1. p. 383.

6. Gold, Thomas. The Origin of Methane (and Oil) in the Crust of the Earth. *United States Geological Survey* Professional Paper No. 1570. 1993.

7. Ibid.

8. Corsi, Jerome. *Coast to Coast AM.* Radio interview. January 12, 2006.

9. Gold. *The Deep Hot Biosphere.* p. 85.

10. Owen, James. Oil Drillers Strike World's Deepest Dinosaur. *National Geographic News.* April 26, 2012.

Chapter 2:

1. US President Jimmy Carter's energy policy speech. April 18, 1977.

2. Lazzaro, Joseph. Another George W. Bush Legacy $150 Oil. *Daily Finance.* August 24, 2009. http://www.dailyfinance.com/2009/08/24/another-legacy-of-president-george-w-bush-150-oil/

3. US President Barack Obama's weekly address. March 3, 2012.

4. Nilsson, Christopher. *Conservation of Energy.* Ontario. GTK Press, 1999. p. 24.

5. Draper, Graham, et al. *Physical Geography 7: Discovering Global System and Patterns.* Ontario. Gage Learning Corporation, 2000. p. 177.

6. DesRivieres, Dennis. *Experience Canada: A Geography*. Ontario. Oxford University Press, 2003. p. 226.

7. Ibid. p. 228.

8. Chasmer, Ron. *Earth Matters: Studies in Physical Geography*. Ontario. Oxford University Press, 2001. p. 31.

9. Gold. *The Deep Hot Biosphere*. p. 5.

10. Tarbuck, Edward J. and Lutgen, Frederick K. *Earth Science 9th edition*. New Jersey. Prentice Hall, 2000. p. 7.

11. Huesemann, Michael. *Why Technology Won't Save Us or the Environment*. British Columbia. New Society Publishers, 2011. p. 21.

12. Total Oil Supply 1980-2011. *US Energy Information Administration*. Available online.

13. Schumach, Murray. Gas Heat Restored as Officials Urge Fuel Saving Here. *The New York Times*. January 27, 1948.

14. Ten More Years of Oil – What Then? *The New York Times*. January 25, 1948.

15. Day, David T. The Petroleum Resources of the United States. *The American Review of Reviews*. 1909.

16. Smith, George Otis. Our Need of New Sources of Power. *Munsey's Magazine*. October 1920, Vol. LXX1. p. 74-75.

17. Look Again. *The Economist*. June 3, 1989. p. 70.

18. Anything But Oil. *The Economist*. May 26, 1990. p. 23.

19. Total Oil Supply 1980-2011. Total Petroleum Consumption 1980-2011. *US Energy Information Administration*. Both available online.

20. Wetuski, Jodi. Experts say the world is running out of gas—and oil. *Oil and Gas Investor*. July 2000. p. 17.

21. Total Oil Supply 1980-2011. *US Energy Information Administration*.

22. Ibid.

23. Eisenbeis, Robert. Are We Running Out of Oil? *EconSouth*. Federal Reserve Bank of Atlanta. Third Quarter 2002. p. 2-4.

24. Tarbell, Ida M. *The History of the Standard Oil Company*. p. 81.

25. Bainerman, Joel. Is the World Running Out of Oil? *The Middle East*. April 2004. p. 6.

26. Annual Motor Gasoline Regular Grade Retail Price. Short-Term Energy Outlook Real and Nominal Prices, September 2012 *US Energy Information Administration*. Available online.

27. Spot Prices for Crude Oil and Petroleum Products. *US Energy Information Administration*. http://www.eia.gov/dnav/pet/pet_pri_spt_s1_m.htm

28. Ibid.

29. Annual Motor Gasoline Regular Grade Retail Price. *US Energy Information Administration*.

30. Deepwater Horizon Drills World's Deepest Oil & Gas Well. *Transocean Ltd.* http://www.deepwater.com/fw/main/IDeepwater-Horizon-i-Drills-Worlds-Deepest-Oil-and-Gas-Well-419C151.html

31. Ibid.

32. Gold. *The Deep Hot Biosphere*. p. 90.

33. DL Annular Blowout Preventer. *Cameron International.* http://www.c-a-m.com/forms/Product.aspx?prodID=ecd6ef4e-19dc-4ea9-ad47-94689e36b61c

34. Hubbert, Marion King. Nuclear Energy and the Fossil Fuels. *Shell Development Company.* June 1956. Publication No. 95. p. 22-26.

35. Total Oil Supply 1980-2011. *US Energy Information Administration.*

36. Ibid.

37. Jackson, Peter M. Why the "Peak Oil" Theory Falls Down: Myths, Legends and the Future of Oil Resources. *Cambridge Energy Research Associates Inc.* November 2006.

38. Ibid.

Chapter 3:

1. Mendeleev, Dmitri. *L'Origine du Petrole.* La Revue Scientifique, No. 18. November 3, 1877. p. 416.

2. Gold. *The Deep Hot Biosphere.* p. vii.

3. Renewable energy has an icon: Henry Ford. *Associated Press.* October 12, 2006. http://www.msnbc.msn.com/id/15233556/ns/us_news-environment/t/renewable-energy-has-icon-henry-ford/

4. Browne, Malcolm W. Geochemist Says Oil Fields May Be Refilled Naturally. *The New York Times.* September 26, 1995.

5. Gold. The Origin of Methane (and Oil) in the Crust of the Earth. *US Geological Survey,* Professional Paper No. 1570, 1993.

6. Creswell, Julie. Oil Without End? Revisionists say oil isn't a fossil fuel. That could mean there's lots more of it. *CNN Money.* February 17, 2003.

7. Watson, Matt. Development and Performance Update of the Cambrian Reservoir. *Merit Energy Company.* The 4th Annual Wyoming CO_2 Conference. June 29-30, 2010.

8. Brokmeyer, R.J., et al. Lost Soldier Tensleep CO_2 Tertiary Project, Performance Case History. Conference Paper for the *Society of Petroleum Engineers.* 35191-MS. March 27-29, 1996.

9. Watson, Matt. Development and Performance Update of the Cambrian Reservoir. June 29-30, 2010.

10. Mahfoud, Robert F. and Beck, James N. Why the Middle East fields may produce oil forever. *Offshore Magazine.* Vol. 55, Iss. 4. April 1995.

11. Titan's surface organics surpass oil reserves on Earth. *European Space Agency.* February 13, 2008.

12. Ibid.

13. Gold. *The Deep Hot Biosphere.* p. 65.

14. Ibid. p. 66.

15. Ibid. p. 67.

16. Franz Fischer and Hans Tropsch, of Mulheim-on-the-Ruhr, Germany. Process for the Production of Paraffin-Hydrocarbons With More Than One Carbon Atom. *United States Patent Office.* Patented February 11, 1930 No. 1,746,464.

17. Lucier, Paul. *Scientists and Swindles: Consulting on Coal and Oil in America, 1820-1890.* Baltimore, MD. *John Hopkins University Press,* 2008. p. 150.

18. Tarbell, Ida M. *The History of the Standard Oil Company.* p.19.

19. Leckel, Dieter. Diesel Production from Fischer-Tropsch: The Past, the Present, and New Concepts. *Energy & Fuels*. March 11, 2009. p. 2342.

20. Wender, Irving and Seshadri, Kalkunte S. et. al. Coal Liquefaction - Investigation of Reactor Performance, Role of Catalysts and PCT Properties. Prepared for *US Department of Energy*. Contract No. DE-FG22-83PC60054. 1983. http://www.fischer-tropsch.org/DOE/DOE_reports/60054/doe_pc_60054-t1/doe_pc_60054-t1.pdf

21. Davis. Burtron H. Technology Development for Fischer-Tropsch Catalysis. Prepared for *US Department of Energy*. Contract No. DE-AC22-94PC94055. Quarterly Technical Progress Report No.12. December 16, 1997.

22. Fischer-Tropsch Fuels. *National Energy Technology Laboratory*. November 2011. http://www.netl.doe.gov

23. Gold. *The Deep Hot Biosphere*. p. 111.

24. Gold. *The Deep Hot Biosphere*. p. 121.

Chapter 4:

1. Bryan, Victoria, et al. Spotlight turns to output as Boeing wins the show. *Reuters*. July 13, 2012.

2. Virgin Atlantic website http://www.virgin-atlantic.com/en/gb/allaboutus/ourfleet/index.jsp

3. Chuter, Andrew. U.K. To Order First Production F-35 for Training. *Defense News*. July 19, 2012. http://www.defensenews.com/article/20120719/DEFREG01/307190001/U-K-Order-First-Production-F-35-Training

4. Assessment of Potential Additions to Conventional Oil and Gas Resources of the World (Outside the United States) from Reserve Growth, 2012. *US Geological Survey.*

5. Arctic oil and natural gas resources. *US Energy Information Administration.* January 20, 2012.

6. Dry Natural Gas Consumption 1980-2011. *US Energy Information Administration.* Available online.

7. Coal Matters: Global Availability of Coal. *World Coal Association*, 2012.

8. Mammals: Elephant. San Diego Zoo's Animal Bytes: Elephant. http://sandiegozoo.org/animalbytes/t-elephant.html

9. The Antonov Giant's: An-22, An-124, & An-225. http://www.airvectors.net/avantgt.html

10. Total Primary Coal Production 1980-2011. *US Energy Information Administration.* Available online.

11. Petroleum Resources Management System. *Society of Petroleum Engineers. American Association of Petroleum Geologists. World Petroleum Council. Society of Petroleum Evaluation Engineers.* 2007.

12. Assessment of Potential Additions to Conventional Oil and Gas Resources of the World, 2012. *US Geological Survey.*

13. Predicts Copper Famine in 20 Years. Bridgeport California, *Chronicle-Union Newspaper.* Estimated publication date July 1924.

Chapter 5:

1. Tarbell, Ida M. *The History of the Standard Oil Company.* p.42.

2. Ibid. p. 87.

3. Ibid. p. 68.

4. Ibid. p. 81.

5. Royal Dutch Petroleum Company Annual Report 2000.

6. Ibid. 2001.

7. Ibid. 2003.

8. Royal Dutch Shell Annual Report 2004.

9. Ibid. 2005.

10. Ibid. 2006.

11. Ibid. 2007.

12. Ibid. 2008.

13. Ibid. 2009.

14. Ibid. 2010.

15. Ibid. 2011.

16. ChevronTexaco Annual Report Supplement 2001.

17. ChevronTexaco Annual Report 2002.

18. Ibid. 2003.

19. Ibid. 2004.

20. Chevron Annual Report 2005.

21. Ibid. 2006.

22. Ibid. 2007.

23. Ibid. 2008.

24. Ibid. 2009.

25. Ibid. 2010.

26. Ibid. 2011.

27. ConocoPhillips SEC Annual Report 2002.

28. ConocoPhillips Annual Report 2003.

29. ConocoPhillips Annual Report Excerpt 2004.

30. ConocoPhillips Annual Report 2005.

31. Ibid. 2006.

32. Ibid. 2007.

33. Ibid. 2008.

34. ConocoPhillips Annual Report Summary 2009.

35. Ibid. 2010.

36. Ibid. 2011.

37. BP Amoco &Arco Combination Shareholders Convention 1999.

38. BP Annual Report & Accounts 2000.

39. BP Annual Report 2001.

40. BP Annual Review 2003.

41. Ibid. 2005.

42. Ibid. 2007.

43. Ibid. 2010.

44. BP Summary Review 2011.

45. ExxonMobil SEC Annual Report 1999.

46. ExxonMobil Summary Annual Report 2003.

47. Ibid. 2004.

48. Ibid. 2006.

49. Ibid. 2007.

50. Ibid. 2008.

51. Ibid. 2009.

52. Ibid. 2010.

53. Ibid. 2011.

54. Annual Motor Gasoline Regular Grade Retail Price. Short-Term Energy Outlook Real and Nominal Prices, September 2012 *US Energy Information Administration.* Available online.

55. Table. H-6. Regions--All Races by Median and Mean Income: 1975-2009. US Census. http://www.census.gov/hhes/www/income/data/historical/household/H06AR_2009.xls

56. *National Lampoons Christmas Vacation.* Warner Bros. Pictures. December 1989.

Chapter 6:

1. *An Inconvenient Truth* documentary. Paramount Classics. May 2006.

2. Akasofu, Syun-Ichi. *The Great Global Warming Swindle* documentary. Martin Durkin. March 2007.

3. Broecker, Wallace S. Are We on the Brink of a Pronounced Global Warming? *Science.* Vol. 189, No. 4201. August 8, 1975.

4. Clark, Ian. *Canadian Energy Senate Committee Hearing.* December 15, 2011.

5. The Encyclopedia of the Earth – Atmospheric Composition. http://www.eoearth.org/article/Atmospheric_composition

6. Brahic, Catherine. Climate myths: Human CO2 emissions are too tiny to matter. *New Scientist.* May 16, 2007.

7. Patterson, Timothy. *Canadian Energy Senate Committee Hearing.* December 15, 2011.

8. Ibid.

9. Hubbert, Marion King. Nuclear Energy and the Fossil Fuels. p. 22-24.

10. Time magazine Global Warming cover. Vol. 167 No. 14. April 3, 2006.

11. Ibid, Vicious Cycles. March 26, 2006.

12. Shikwati, James *The Great Global Warming Swindle.*

13. US President Barrack Obama's Speech and Interaction on the Young African Leader's Initiative in South Africa, Uganda, Nigeria and Kenya. *NTV Uganda.* Live stream: June 29, 2013. http://www.youtube.com/watch?v=03Qdh7CdzjA

14. Calder, Nigel *The Great Global Warming Swindle.*

15. Goldenberg, Suzanne. Barack Obama swipes at Mitt Romney over climate change jibe. *The Guardian.* September 7, 2012. http://www.guardian.co.uk/environment/2012/sep/07/barack-obama-mitt-romney-climate

16. United Nations Environment Programme Annual Report 2003.

17. Ibid. 2004.

18. Ibid. 2005.

19. Ibid. 2006.

20. Ibid. 2007.

21. United Nations Environment Programme Overview 2010-2011 & 2008-2009.

22. United Nations Environment Programme 2009 Annual Report.

23. Ibid. 2010.

24. Ibid. 2011.

25. Canada's International Action. http://www.climatechange.gc.ca/default.asp?lang=En&n=5497F282-1

26. A Climate Change Plan for the Purposes of the Kyoto Protocol Implementation Act – May 2010. http://www.climatechange.gc.ca/default.asp?lang=En&n=3DA45D49-1

27. Annual Report to Congress: Our Changing Planet. http://www.climatescience.gov

28. Federal Climate Change Programs: Funding History and Policy Issues. *Congressional Budget Office* study. March 2010.

29. Federal Climate Change Expenditures Report to Congress. June 2010. http://www.whitehouse.gov/sites/default/files/omb/assets/legislative_reports/FY2011_Climate_Change.pdf

30. Lattanzio, Richard K. The Global Climate Change Initiative (GCCI): Budget Authority and Request, FY2010-FY2013. *Congressional Research Service*. July 27, 2012.

31. UK International Climate Fund: Tackling climate change, reducing poverty. *Department for International Development. Department of Energy & Climate Change. Department for Environment Food and Rural Affairs.*

32. President's Review & Annual Report 1988. The Rockefeller Foundation.

33. Ibid. 1989.

34. Ibid. 1991.

35. Ibid. 1994.

36. Grants and Grantees: Developing Climate Change Resilience. The Rockefeller Foundation. http://www.rockerfellerfoundation.org/grants/search?initiativeIds=1&limit=40

37. Committee on Energy and Natural Resources U.S. Senate, 100th Congress 1st session on the greenhouse effect and global climate change. Max Baucus speech. June 23, 1988 Part 2.

38. *Swordfish.* Warner Bros. Pictures. June 2001.

Chapter 7:

1. Davidson Fellow William Yuan. *The Davidson Institute for Talent Development.*

2. Harding, Susan. Beaverton boy lauded for solar cell invention. *KATU.* September 15, 2008. http://katu.com/news/2843298.html

3. Hicks, Jennifer. Grass Clippings Turned Into Solar Cells by MIT Scientist. *Forbes.* February 21, 2012.

Chapter 8:

1. Science Content Standards for California Public Schools Kindergarten Through Grade Twelve. *California Department of Education*, 2000.

2. Total Oil Supply 1980-2011. *US Energy Information Administration*. Available Online.

3. Is rolling back royalty rates the right move in Alberta? *CBC News*. March 12, 2010.

4. Goodman, Leah McGrath. The gas tax you don't even know you're paying. *CNN Money*. April 23, 2012.

5. Ontario Ministry of Transportation History. http://www.mto.gov.on.ca/english/about/history.shtml

6. Bickley, James M. The Federal Excise Tax on Gasoline and the Highway Trust Fund: A Short History. *Congressional Research Service*. March 9, 2012.

Chapter 9:

1. Gold. *The Deep Hot Biosphere*. p. 90.

2. Ibid. p. 125-126.

3. Ibid. p. 28.

4. Ibid. p. 101.

5. Ibid. p. 100-103.

6. General Electric to Pay $234,000 for Environmental Violations. *New York State News Department of Environmental Conversation*. February 3, 1998.

7. General Electric Annual Report Financial Section 2000.

8. Ibid. 2011.

Final Thoughts:

1. Moreau, Ron & Yousafzai, Sami. Can Buried Treasure Save Afghanistan? *The Newsweek Daily Beast Company*. June 13, 2010.

2. Damail, Jahr. Western oil firms remain as US exits Iraq. *Al Jazeera English*. January 7, 2012.

3. Fisher, Max. Why it's good news for the U.S. that China is snapping up Iraq's oil. *The Washington Post*. June 3, 2013.

4. Ibid.

5. Changfoot, Nadine. Feminist Standpoint Theory, Hegel and the Dialectical Self: Shifting the Foundations. *Philosophy & Social Criticism*, Vol. 30. June 2004. p. 478.

6. Harding, Sandra. A Socially Relevant Philosophy of Science? Resources from Standpoint Theory's Controversiality. *Hypatia*, Volume 19, No. 1. Winter 2004. p. 30.

Selected Bibliography

Alekperov, Vagit. *Oil of Russia: Past, Present & Future.* Minneapolis, MN. Eastview Press, 2011.

Bainerman, Joel. Is the World Running Out of Oil? *The Middle East.* April 2004.

Bartlett, Bruce. Oil unlimited? *The Washington Times.* Tuesday, June 8, 2004.

Bell, Larry. *Climate of Corruption: Politics and Power Behind the Global Warming Hoax.* Austin, TX. Greenleaf Book Group Press, 2011.

Bickley, James M. The Federal Excise Tax on Gasoline and the Highway Trust Fund: A Short History. *Congressional Research Service.* March 9, 2012.

Brahic, Catherine. Climate myths: Human CO2 emissions are too tiny to matter. *New Scientist.* May 16, 2007.

Broecker, Wallace S. Are We on the Brink of a Pronounced Global Warming? *Science.* Vol. 189, No. 4201. August 8, 1975.

Brokmeyer, R. J., et al. Lost Soldier Tensleep CO_2 Tertiary Project, Performance Case History. Conference Paper for the *Society of Petroleum Engineers*. 35191-MS. March 27-29, 1996.

Browne, Malcolm W. Geochemist Says Oil Fields May Be Refilled Naturally. *The New York Times*. September 26, 1995.

Bryan, Victoria, et al. Spotlight turns to output as Boeing wins the show. *Reuters*. July 13, 2012.

Butorac, Mark. From The Other Oil Field: Mendeleev, the West and the Russian Oil Industry. Montreal, QC. McGill University Department of History, July 2001.

Changfoot, Nadine. Feminist Standpoint Theory, Hegel and the Dialectical Self: Shifting the Foundations. *Philosophy & Social Criticism*, Vol. 30. June 2004.

Chasmer, Ron. *Earth Matters: Studies in Physical Geography*. Don Mills, ON. Oxford University Press, 2001.

Chuter, Andrew. U.K. To Order First Production F-35 for Training. *Defense News*. July 19, 2012.

Creswell, Julie. Oil Without End? Revisionists say oil isn't a fossil fuel. That could mean there's lots more of it. *CNN Money*. February 17, 2003.

Damail, Jahr. Western oil firms remain as US exits Iraq. *Al Jazeera English*. January 7, 2012.

Davis. Burtron H. Technology Development for Fischer-Tropsch Catalysis. Prepared for *US Department of Energy*. Contract No. DE-AC22-94PC94055. Quarterly Technical Progress Report No. 12. December 16, 1997.

Day, David T. The Petroleum Resources of the United States. The American Review of Reviews. 1909.

DesRivieres, Dennis. *Experience Canada: A Geography*. Don Mills, ON. Oxford University Press, 2003.

Draper, Graham, et al. *Physical Geography 7: Discovering Global System and Patterns*. Scarborough, ON. Gage Learning Corporation, 2000.

Eisenbeis, Robert. Are We Running Out of Oil? *EconSouth*. Federal Reserve Bank of Atlanta. Third Quarter 2002.

Fisher, Max. Why it's good news for the U.S. that China is snapping up Iraq's oil. *The Washington Post*. June 3, 2013.

Gold, Thomas. *The Deep Hot Biosphere: The Myth of Fossil Fuels*. Ney York, NY. Copernicus Books, 2001.

—. The Origin of Methane (and Oil) in the Crust of the Earth. *US Geological Survey* Professional Paper No. 1570. 1993.

Goldenberg, Suzanne. Barack Obama swipes at Mitt Romney over climate change jibe. *The Guardian*. September 7, 2012.

Goodman, Leah McGrath. The gas tax you don't even know you're paying. *CNN Money*. April 23, 2012.

Harding, Sandra. A Socially Relevant Philosophy of Science? Resources from Standpoint Theory's Controversiality. *Hypatia*, Volume 19, No. 1, Winter 2004.

Harding, Susan. Beaverton boy lauded for solar cell invention. *KATU*. September 15, 2008.

Hicks, Jennifer. Grass Clippings Turned Into Solar Cells by MIT Scientist. *Forbes*. February 21, 2012.

Hubbert, Marion King. Nuclear Energy and the Fossil Fuels. *Shell Development Company*. June 1956. Publication No. 95.

Huesemann, Michael. *Why Technology Won't Save Us or the Environment*. Gabriola, BC. New Society Publishers, 2011.

Jackson, Peter M. Why the "Peak Oil" Theory Falls Down: Myths, Legends and the Future of Oil Resources. *Cambridge Energy Research Associates Inc.* November 2006.

Lattanzio, Richard K. The Global Climate Change Initiative (GCCI): Budget Authority and Request, FY2010-FY2013. Congressional Research Service. July 27, 2012.

Lazzaro, Joseph. Another George W. Bush Legacy $150 Oil. *Daily Finance.* August 24, 2009.

Leckel, Dieter. Diesel Production from Fischer-Tropsch: The Past, the Present, and New Concepts. *Energy & Fuels.* March 11, 2009.

Lucier, Paul. *Scientists and Swindles: Consulting on Coal and Oil in America, 1820-1890.* Baltimore, MD. *John Hopkins University Press*, 2008.

Mahfoud, Robert F. and Beck, James N. Why the Middle East fields may produce oil forever. *Offshore Magazine.* Vol. 55, Iss. 4. April 1995.

Mendeleev, Dmitri. *L'Origine du Petrole.* La Revue Scientifique, No. 18. November 3, 1877.

Moreau, Ron & Yousafzai, Sami. Can Buried Treasure Save Afghanistan? *The Newsweek Daily Beast Company.* June 13, 2010.

Neumann, Casper. *The Chemical Works of Caspar Neumann*, M. D. Trans. & ed. William Lewis. London, 1759. 2nd edition 1773. Vol. 1.

Nilsson, Christopher. *Conservation of Energy.* Toronto, ON. GTK Press, 1999.

Olien, Diana Davids and Olien, Roger M. Running Out of Oil: Discourse and Public Policy, 1909-1929. *The University of Texas-Permian Basin, Department of History*, 1993.

Owen, James. Oil Drillers Strike World's Deepest Dinosaur. *National Geographic News*. April 26, 2012.

Schumach, Murray. Gas Heat Restored as Officials Urge Fuel Saving Here. *The New York Times*. January 27, 1948.

Smith, George Otis. Our Need of New Sources of Power. *Munsey's Magazine*. October 1920. Vol. LXX1.

Tarbell, Ida M. *The History of the Standard Oil Company*. New York, NY. McClure, Phillips & Company, 1904.

Tarbuck, Edward J. and Lutgen, Frederick K. *Earth Science 9th edition*. Upper Saddle River, NJ. Prentice Hall, 2000.

Watson, Matt. Development and Performance Update of the Cambrian Reservoir. *Merit Energy Company*. The 4th Annual Wyoming CO_2 Conference. June 29-30, 2010.

Wender, Irving and Seshadri, Kalkunte S. et. al. Coal Liquefaction - Investigation of Reactor Performance, Role of Catalysts and PCT Properties. Prepared for *US Department of Energy*. Contract No. DE-FG22-83PC60054. 1983.

Wetuski, Jodi. Experts say the world is running out of gas—and oil. *Oil and Gas Investor*. July 2000.

Index

A

B

C

D

E

F

G

H

I

K

L

M

N

O

P

R

S

Made in United States
Orlando, FL
11 November 2026

90609405R00102